Domain Structured Dynamics

Unpredictability, chaos, randomness, fractals, differential equations and neural networks

Domain Structured Dynamics

Unpredictability, chaos, randomness, fractals, differential equations and neural networks

Marat Akhmet
Department of Mathematics, Middle East Technical University, Ankara, Turkey

IOP Publishing, Bristol, UK

ISBN 978-0-7503-3507-2 (ebook)
ISBN 978-0-7503-3505-8 (print)
ISBN 978-0-7503-3508-9 (myPrint)
ISBN 978-0-7503-3506-5 (mobi)

DOI 10.1088/978-0-7503-3507-2

Version: 20210801

IOP ebooks

British Library Cataloguing-in-Publication Data: A catalogue record for this book is available from the British Library.

Published by IOP Publishing, wholly owned by The Institute of Physics, London

IOP Publishing, Temple Circus, Temple Way, Bristol, BS1 6HG, UK

US Office: IOP Publishing, Inc., 190 North Independence Mall West, Suite 601, Philadelphia, PA 19106, USA

To my beloved family.

Contents

Part III Random processes

Part IV Non-autonomous differential equations

Part V Neural networks

Preface

The book has been designed as the simplest way to understand mathematical chaos as well as to open a gate to a deeper perception of the world in motion. A new modeling tool has been invented for this task, the *abstract similarity dynamics*. The product of the application of the tool is called the *domain structured dynamics*. Accordingly, the *labeling* for state space points is suggested and the result is called the *abstract similarity set*. The *abstract similarity map* is another novelty, coupled with the labeling to complete the toolkit for the research. We strongly believe that the domain structuring and the abstract similarity map are effective instruments to investigate many problems of modern dynamics, chaos first of all. The results of this book serve to confirm this belief. In terms of presentation we follow the confidence exhibited by Stephen Hawking in [25]: 'Scientists and engineers tend to express their ideas in the form of equations because they need to know the precise value of quantities. But for the rest of us, a qualitative grasp of scientific concepts is sufficient, and this can be conveyed by words and diagrams, without the use of equations.' A result of the present research activity is the conviction that the labeling which has been invented in our studies is more picturesque than any equations and methods considered for the same goals. Thus, the structure will help researchers to understand chaos more easily and the power of domain structured dynamics is guaranteed by the theoretical depth of our research. Practically, it is determined by the existing and potential algorithms of labeling. The most familiar one is the Fatou–Julia iteration, which is applied for self-similarity structures, fractals, and chaos in discrete and differential equations and neural networks in our study. It is clear that the products of artificial intelligence can be utilized to realize the idea for industrial applications. Currently, this could be deep learning analysis, if designed properly. Our suggestion to pay more attention to the geometrical problems of chaos is also productive. In [29] Nicolis and Prigogine claim that 'It is more natural, or at least less ambiguous to speak of *complex behavior* rather than complex systems. The study of such behavior will reveal certain common characteristics among different classes of systems and will allow us to arrive at a proper understanding of complexity.' What we are doing is the investigation of complexity without ignoring the domains of chaos. For example, we have found that systems with complex behavior can be significantly simple. Additionally, a very common characteristic, for the complexity of behavior has been identified. It is the abstract similarity map and now it is a challenging problem to show that the map can be universal for complex dynamics. That is, a consequence of our research may be that the complexity of behavior is somewhat of an illusion.

We provide theories of two types of chaos, which have been developed in our recent papers. The first is the *domain structured chaos* [1–3, 5, 11, 16] and the second is the *unpredictable dynamics* [7, 8].

The domain structured chaos is a specific type of behavior for *domain structured dynamics*. Theoretically, the *abstract similarity dynamics* lies at the basis of the domain structured dynamics, which occurs in *abstract similarity sets* through the *abstract similarity map* and can be exemplified by symbolic dynamics, but most formally, as it relates to symbolic dynamics as integration to summation, in calculus. That is, the symbolic dynamics is a special case of the abstract similarity dynamics when the labeling for the set of symbol sequences is maximally simple. Our proposals surpass the symbolic dynamics also in the methodological sense. To study motion through symbolic dynamics one has to project it on the Bernoulli shift along a sequence of symbols. The labeling is an isometry which is determined by the process. Thus, we avoid a search for the projection's characteristics and solve a problem completely inside the state space. We are confident that the new method can be utilized as a universal method for the study of the most sophisticated dynamics as well as a method for chaos initiation in many applications. The reader will see that domain structured chaos has already been applied to the extension of chaos in hyperbolic sets, chaotic attractors, fractals, finite dimensional cubes, stochastic dynamics, probability, differential equations and neural networks in this book on the basis of the papers [1–5, 9, 10, 12–14, 16, 19, 22].

The cornerstone of the newly introduced unpredictable dynamics is the *unpredictable point*, which is a specification of the Poisson stable point. Let us recall that the Poisson stable point is known as the ultimate type of recurrence in the Poincaré–Birkhoff theory [23, 28, 31, 32]. Thus, our suggestions prolong the line of recurrent motions to a new frontier such that the research joins with chaos investigation. It is important to emphasize that Devaney's chaos relies completely on the collective behavior of orbits such as sensitivity, dense cycles and transitivity [24, 27]. In contrast, Poincaré chaos [7, 8] is based on the unpredictability concept, which is focused uniquely on a single orbit motion. Moreover, by simulating a solution we are able to make conclusions not only on the presence of chaos, but also that the motion is unpredictable [15]. This is, on one hand, principally new for the study of chaos and on the other hand leads us back to the roots of classical dynamics, which considers a single stable motion as the main object of discussion. What we are doing is akin to the idea of the one-electron universe. In our opinion the novel motion will make the research of chaos more beneficial within theoretical research as well as for applications. The first discoveries have already been reported in papers [1, 2], where we have deepened the understanding of random processes through the two types of chaos invented for our research. In the current book the reader will find many interesting models being revisited. To shed more light on the concept of unpredictability, let us say, similarly to the title of [27], that *unpredictability implies chaos*. However, the unpredictability is of much weaker recurrence than periodicity and the results in [8] and in [27] make the boundary points of the spectrum such that all other types of recurrence, quasi-periodicity, almost periodicity and others, can be found inside it. That is, we simply continue to look for what one might call the *order in chaos*.

To study stochastic dynamics and probability means to find *necessary* features of the process. Deterministic chaos is formally a list of conditions which *necessarily* must happen to say that the phenomenon is present in a model. Consequently, one can search for some ingredients of chaos in the dynamics and the reader must be confident that we do not abolish randomness through this study. This research would be easier if the conditions for chaos are more fit to what is waiting, and this is what we are working on through the introduction of the new methods of chaos investigation. The results of our research activity are presented in this book. For example, we have shown that unpredictable dynamics necessarily occur in Bernoulli schemes and Markov chains. The relations of chaos and random processes are the focus of many studies [26, 30]. It should be emphasized that specifically Poincaré chaos plays the most important role in the research and not the conventional types of chaos.

Our proposals are very effective in the study of chaos and random processes, and the future development of science will demonstrate this. At the moment, let us mention the most simple examples to emphasize not the novelty, but the effectiveness of the method. Domain structured chaos is convenient for the analysis of logistic maps and baker's maps. Abstract hyperbolic chaos is verified optimally for the Smale horseshoe and baker two-dimensional transformation. The reader is invited to compare our suggestions with the results which are known from traditional methods. The new approach of labeling is very helpful for exploring chaos in fractals. The episodic cases of chaos indication in the Cantor set or Sierpinski carpet are now replaced by the universal method for fractals [5]. Moreover, surprisingly, we have explored how to arrange chaotic dynamics in sets which are not fractals but finite sets and finite dimensional cubes and, consequently, to all those which are topologically equivalent to them. The concept of abstract fractals has been introduced [16], which is helpful for modern science. A difficult and challenging problem for recurrent motions is the functional analysis of time series. The most developed, in this sense, are periodic functions. Unfortunately, the analytical presentation of almost periodic functions is already an unsolvable problem. Although there are very particular examples of quasi-periodic functions, there are no developed examples of Poisson stable functions. In the current research we consider the problem for unpredictable functions by presenting them as solutions of difference and differential equations as well as functions determined randomly. The results are in use for differential equations and neural networks as inputs and outputs. We investigate neural networks with unpredictable oscillations in papers [6, 17, 18, 20, 21]. The models are indisputably important for neuroscience and artificial intelligence problems. Dynamics with irregular behavior reflects the complexity and effectiveness of the human mind and significantly support the power and universality of computers and robots.

References

[1] Akhmet M 2019 Domain structured chaos for discrete random processes arXiv:1912.10478 [math.DS]

[2] Akhmet M 2020 Domain structured chaos in Markov chains arXiv: 2010.14625v1 [math.DS]

[3] Akhmet M and Alejaily E M 2019 Domain-structured chaos in a Hopfield neural network *Int. J. Bifurcat. Chaos* **29** 1950205

[4] Akhmet M and Alejaily E M 2019 Finite dimensional space chaotification *Kazakh Math. J.* **19** 21–6

[5] Akhmet M and Alejaily E M 2021 Abstract similarity, chaos and fractals *Discrete Contin. Dyn. Syst. Ser.* B **26** 2479–97
[6] Akhmet M, Aruğaslan Çinçin D, Tleubergenova M and Nugayeva Z 2021 Unpredictable oscillations for Hopfield-type neural networks with delayed and advanced arguments *Mathematics* **9** 1–19
[7] Akhmet M and Fen M O 2016 Poincaré chaos and unpredictable functions *Commun. Nonlinear Sci. Numer. Simul.* **48** 85–94
[8] Akhmet M and Fen M O 2016 Unpredictable points and chaos *Commun. Nonlinear Sci. Numer. Simul.* **40** 1–5
[9] Akhmet M and Fen M O 2017 Existence of unpredictable solutions and chaos *Turk. J. Math.* **41** 254–66
[10] Akhmet M and Fen M O 2018 Non-autonomous equations with unpredictable solutions *Commun. Nonlinear Sci. Numer. Simul.* **59** 657–70
[11] Akhmet M, Fen M O and Alejaily E M 2018 Dynamics motivated by Sierpinski fractals arXiv: 1811.07122
[12] Akhmet M, Fen M O, Tleubergenova M and Zhamanshin A 2019 Unpredictable solutions of linear differential and discrete equations *Turk. J. Math.* **43** 2377–89
[13] Akhmet M, Fen M O, Tleubergenova M and Nugayeva Z 2020 Unpredictable solutions of linear impulsive systems *Mathematics* **8** 1798
[14] Akhmet M, Fen M O, Tleubergenova M and Zhamanshin A 2019 Poincaré chaos for a hyperbolic quasilinear system *Miskolc Math. Notes* **20** 33–44
[15] Akhmet M, Fen M O and Tola A 2021 A numerical analysis of Poincaré chaos *Discontinuity Nonlinearity Complex* (in press)
[16] Akhmet M and Alejaily E M 2021 Abstract fractals *Discontin. Nonlinear Complex* **10** 135–42
[17] Akhmet M, Seilova R, Tleubergenova M and Zhamanshin A 2020 Shunting inhibitory cellular neural networks with strongly unpredictable oscillations *Commun. Nonlinear Sci. Numer. Simul.* **89** 105287
[18] Akhmet M, Tleubergenova M and Nugayeva Z 2020 Strongly unpredictable oscillations of Hopfield-type neural networks *Mathematics* **8** 1791
[19] Akhmet M, Tleubergenova M and Zhamanshin A 2020 Quasilinear differential equations with strongly unpredictable solutions *Carpathian J. Math.* **36** 341–9
[20] Akhmet M, Tleubergenova M and Zhamanshin A 2019 Neural networks with Poincaré chaos *ACM Int. Conf. Proc. Ser.* pp 34–7
[21] Akhmet M, Tleubergenova M and Zhamanshin A 2020 Inertial neural networks with unpredictable oscillations *Mathematics* **8** 1797
[22] Akhmet M and Tola A 2020 Unpredictable strings *Kazakh Math. J.* **20** 16–22
[23] Birkhoff G D 1991 *Dynamical Systems* (Providence, RI: Colloquium)
[24] Devaney R L 1989 *An Introduction to Chaotic Dynamical Systems* (Canada; USA: Addison-Wesley)
[25] Hawking S 1994 *Black Holes and Baby Universes* (London: Bantam)
[26] Lasota A and Mackey M C 1985 *Probabilistic Properties of Deterministic Systems* (Cambridge: Cambridge University Press)
[27] Li T Y and Yorke J A 1975 Period three implies chaos *Am. Math. Mon.* **82** 985–92
[28] Nemytskii V V and Stepanov V V 1960 *Qualitative Theory of Differential Equations* (Princeton, NJ: Princeton University Press)

[29] Nicolis G and Prigogine I 1989 *Exploring Complexity: An Introduction* (New York: Freeman)

[30] Ornstein D 1970 Bernoulli shifts with the same entropy are isomorphic *Adv. Math.* **4** 337–52

[31] Poincaré H 1957 *New Methods of Celestial Mechanics* vol 1–3 (New York: Dover)

[32] Sell G R 1971 *Topological Dynamics and Ordinary Differential Equations* (London: Van Nostrand Reinhold)

Acknowledgements

I am very grateful to numerous colleagues who have supported me in my research journey, with special thanks to Ejaily Alejaily. Our productive discussions on fractals and chaos for fractals resulted in the core of this book: abstract similarity and domain structured dynamics. His picturesque simulations encouraged us to look for unexpected events in the world of sophisticated motions and geometries. The concept of unpredictability, or deterministic unpredictability to be more accurate, is an indisputable ingredient of the domain structured chaos and it was initiated and developed in collaboration with Mehmet Onur Fen, to whom I express my kindest gratitude. I would also like to thank the very friendly Astrit Tola for results on unpredictability in random processes, in particular for the laws of large strings, Madina Tleubergenova, Roza Sejilova, Duygu Arugaslan, Akylbek Zhamanshin and Zakhira Nugaeva for their papers, which are the cornerstones of the theory of unpredictable oscillations and applications in neuroscience. I strongly believe that we and all who join us shall extend these concepts as far as needed in future research.

I am greatly indebted to Ashley Gasque Stec, Senior Commissioning Editor at IOP Publishing, for the invitation to write this book and her professionalism, and Robert Trevelyan, Editorial Assistant at IOP Publishing, for help in shaping the manuscript.

The author has been supported by 2247—A National Leading Researchers Program of TUBITAK, Turkey, N 120C138.

Author biography

Marat Akhmet

Dr Marat Akhmet is a Professor at the Department of Mathematics, Middle East Technical University, Turkey. He received his BSc in mathematics from Aktobe State University, Kazakhstan, and his MSc and PhD in differential equations and mathematical physics at Kiev State University, Ukraine. He has been awarded the Science Prize of TUBITAK (Turkey, 2015) for notable achievements in scientific research.

Professor Akhmet is the author of seven books: *Principles of Discontinuous Dynamical Systems* (Springer, 2010), *Nonlinear Hybrid Continuous Discrete-Time Models* (Atlantis Press, 2011), *Neural Networks with Discontinuous Impact Activations* (Springer, 2014), *Replication of Chaos in Neural Networks, Economics and Physics* (Springer, 2015), *Bifurcation in Autonomous and Nonautonomous Differential Equations with Discontinuities* (Springer, 2017), *Dynamics with Chaos and Fractals* (Springer, 2020) and *Almost Periodicity, Chaos and Asymptotical Equivalence* (Springer, 2020).

He has contributed to the fundamentals of the theory of discontinuous almost periodic functions and the solutions of impulsive systems, and has introduced and developed the *B*-equivalence method for systems with variable moments of impulses. The method became effective in the research of models with discontinuities, depending on the parameters. Most sophisticated are those with bifurcation problems and, ultimately, discontinuous dynamics with grazing phenomena, which are of the most interest in mechanics. In recent papers the author presented a new approach to the phenomena, suggested on the basis of singularity, when various effects compensate each other to produce regular dynamics. Significant results have been obtained by Professor Akhmet for differential equations with a piecewise constant argument. The approach of reduction to discrete equations, previously unique, has been replaced by the method of equivalent integral equations. Thus the classical results of operator theory are now in use in research and a new type of generalized dependence on piecewise constant argument has been introduced. Professor Akhmet solved the second Peskin conjecture for integrate-and-fire biological oscillators through a skillful formalization of the firing connection to achieve synchronization. This makes possible the elaboration of the results for the first Peskin conjecture considered previously. There are still many interesting problems of synchronization for biological oscillators, first of all neural networks, which can be investigated using the method. The asymptotic equivalence of equations is a powerful tool of qualitative theory and the author has achieved very subtle conditions for the equivalence, considering special transformations of linear systems. The class of equivalent systems has been essentially enlarged. Fractals are indebted to dynamics, not only being constructed by iterations, but by being states and orbits. This was, for the first time in the literature, demonstrated in the author's research.

Thus the beauty of fractals can now also be extended along the time axis. Chaos is a very attractive object for study and it still relates to the qualitative theory of differential equations in a very narrow spectrum. Primarily numerical algorithms are in use. The huge theoretical background accumulated for differential equations does not deserve to be ignored in chaos research and Professor Akhmet has suggested the replication of the chaos method. The core of the method is the input–output mechanism, which keeps all chaotic ingredients known in the literature under certain conditions on differential and discrete equations. The mechanism has been realized for quasi-linear systems, but it admits great potential for other methods of nonlinear analysis. Chaos results belong to two lines of research: the first was initiated by Poincaré in the depths of the classical theory of dynamical systems and the other is the approach of Lorenz, who introduced the concept of sensitivity, which was discovered numerically. The sensitivity is convenient to look for results in more engineering based problems. The author suggested extending the types of recurrent motion, such as periodic, almost periodic, recurrent and Poisson stable, with the specific concept of the unpredictable point. This brings the research of chaos back in line with Poincaré and provides ways how the Lorenz chaos is further elaborated. The full information on this unpredictability is presented in this book. Likewise, this book includes the complete comprehension of the abstract similarity dynamics and domain structured dynamics, including abstract similarity and domain structured chaos, which are the latest results of Professor Akhmet.

Part I

Introduction

Chapter 1

What this book is about

This book is about very recently established avenues in the development of chaos theory. They are not familiar to the majority of the scientific community. This is why, first of all, we should shed light on the shared aspects of the new results with the former achievements of the theory. In addition, the differences have to be stressed and, importantly, the advantages of the new opportunities for research should be clarified. All these are considered in the current chapter. Section by section we will consider the unpredictability, abstract similarity chaos, domain structured chaos generated by the abstract similarity map, randomness and fractals. In the first section one of the most important ingredients of the domain structured chaos, unpredictability, is discussed, to emphasize the integrity of the content of this book. The second section provides characteristics for the abstract similarity chaos and domain structured chaos, and provides arguments as to why our proposals maximally speed up the process of chaos indication and are effective for chaos construction in deterministic dynamics. Similar problems are described in the subsequent section, but for stochastic processes. The related topics for fractals are considered in the fourth section of this chapter. They rely on the domain structuring and reflect the problem of chaos arrangement in fractals as well as abstract fractals. The last two sections discuss the development of the theory of differential equations and oscillations with unpredictability in neuroscience.

1.1 Unpredictability

A large part of this book concerns the concept of the unpredictability, which was introduced in our paper [9]. The concept is a result of the intersection of two lines of chaos research. The first stems from and lies within the Poincaré–Birkhoff theory [31, 84]. That is, the *Poincaré line* or *recurrence line*. One can also call it the oscillation line. Another stems from the Lorenz sensitivity [68] and accordingly we call it the *Lorenz line* or *sensitivity line*. Correspondingly, one can speak of *Poincaré*

doi:10.1088/978-0-7503-3507-2ch1

chaos and *Lorenz chaos*. If the second unites all results with sensitivity, Li–Yorke chaos is included, while the first one relies only on the unpredictability.

We start with the definition of the unpredictable point. Let (X, d) be a metric space and assume that f is a flow on X with continuous or discrete time.

Definition 1.1.1. [9] *A point $p \in X$ and the trajectory through it are unpredictable if there exist a positive number ϵ_0 (the unpredictability constant) and sequences $\{t_n\}$ and $\{\tau_n\}$, both of which diverge to infinity, such that*

- $\lim_{n\to\infty} f(t_n, p) = p$;
- $d[f(t_n + \tau_n, p), f(\tau_n, p)] \geqslant \epsilon_0$ for each $n \in \mathbb{N}$.

The behavior of the sequence $f(t_n, p)$ in the last definition is known as the *Poisson stability* and illustrates very clearly what researchers call a recurrent motion. The divergence estimated by ϵ_0 is said to be the *separation property*. The most simple recurrent motion is a fixed point, when the point returns to its origin state at each moment of time. The Poisson stability is the most sophisticated type of recurrence. Others are simpler: periodic, quasi-periodic, almost periodic and recurrent motions [79, 88]. One can say that recurrence is weakened for the types of oscillations starting from a fixed point to the Poisson stable point. The latter is the ultimate point in the Birkhoff classification [31]. Correspondingly, the stability was considered as sufficient to express properly the complexity of celestial mechanics [84]. Some authors accept the dynamics on the quasi-minimal sets with Poisson stable orbits as chaotic. We do not agree that this is true, since there is yet no sensitivity. In our research [8, 9] we have prolonged the line of recurrent motions to unpredictable motions through definition 1.1.1, such that the most weak version of recurrent dynamics has been formulated (we are not confident that the line cannot be prolonged further). More precisely, we specified the Poisson stable point to the unpredictable point. The property of unpredictability guarantees the Lorenz sensitivity [9] and, consequently, now one can say that there is chaos. We called it Poincaré chaos in [9], for the first time in the literature. In figure 1.1 one can see from the Venn diagram how the different types of recurrent motions relate to each other. We will proceed the discussion after the following presentation of the *Lorenz line*, which is the second line in chaos research. The American meteorologists Lorenz observed sensitivity in his numerical experiments on weather prediction [68]. That is, he discovered that small changes in initial conditions produce large changes in long-term outcomes. Next he provided an amazing explanation of the phenomenon by considering a special one-dimensional map derived from the model. Thus deterministic chaos was discovered for the first time. One can say that it was the birth of the chaos epoch, that is of the *Lorenz chaos*.

We say that the unpredictability is at the intersection of the two lines, but this does not mean that there are no other intersection points for the lines. One can remember the dense periodic motions in the dynamics of both lines and the existence of a dense trajectory in the domain of chaos [39] and in the quasi-minimal set [31]. The latter set is the closure of the Poisson stable trajectory [79] as well as the

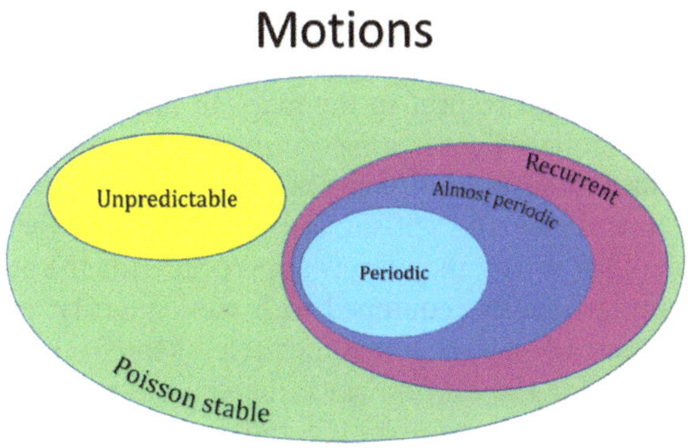

Figure 1.1. The diagram illustrates the most important motions of the Poincaré–Birkhoff theory extended with unpredictable motions, which were introduced in [9].

unpredictable trajectory [9]. Sensitivity is the property which is basic for chaos and it has not been articulated for the quasi-minimal set [88]. Therefore, it would be interesting to find a property for the dynamics in the quasi-minimal set, which plays a similar role and generates the same features as similarity does. This property was discovered in our research. The concept of the unpredictability is a counterpart to the sensitivity. Remarkably, sensitivity cannot be a consequence of the Poisson stability. The unpredictability [9] is the additional condition which creates the new type of dynamics. We proved in our paper that the unpredictability of a single trajectory implies sensitivity in the quasi-minimal set. Thus, the new Poincaré chaos was shaped and the intersection of the Poincaré and Lorenz lines has been completed. This brings the study of Lorenz chaos back to the Poincaré–Birkhoff theory and creates new opportunities for both lines of research. Moreover, we can say that the investigation of chaos can now be performed in the framework of the dynamics of single motions, as is true for stable oscillations in dynamical systems and differential equations [78]. In this direction, the results of papers [10, 11, 13–15, 19, 20] have already been obtained.

1.2 Domain structured dynamics

Our approach to chaos research is characterized by the priority of the domain over the map or the vector field [24]. While the traditional definition of chaos starts with the description of a map with certain properties such as unimodality, hyperbolicity, period three and topological conjugacy to a standard chaotic map such as Bernoulli shift. For example, for the Li–Yorke chaos, the map has to be first defined such that it admits the period three property and then the domain (scrambled set) for the chaos is shaped with specific properties. The same can be said for the chaos of unimodal maps, when the domain is a Cantor set. In the Smale horseshoe case, one can observe that the map and the domain are simultaneously determined so that the construction of the domain starts with an initial set and is shaped step by step using a particular map. Therefore, the structure of the domain and the nature of the map are mutually dependent on each other. For chaos in symbolic dynamics, the domain is primarily described as infinite sequences of symbols and the map is then introduced as a shift on the space. However, the map still has priority, since the properties of the sequences are described with respect to the map. In the current approach, we first construct an abstract similarity set in a metric space with specific conditions to be a suitable venue for the manifestations of chaos. Thereafter, the similarity map is built on the basis of the invariance and self-similarity properties of the domain to obtain an abstract motion. This is why the map is appropriate for the abstract self-similar set as well as for any geometries equipped with self-similarity, and thence proving chaos for these classes of sets becomes possible. Moreover, one can drop the continuity requirement for the motion since the chaotic map does not need to be continuous [1, 61]. In [61], for example, the authors ignore the continuity of some chaotic maps during the discussion of chaos conditions on the product of semi-flows. We regard the continuity of a chaotic map as an important property only from the analytical side, that is to say, it is very useful for handling the map to prove the presence of chaos [100], however, it is not rigorously correct to consider it as an intrinsic property of chaos.

Symbolic dynamics is a generic phenomenon in mathematics. To follow a dynamics with continuous time precisely one needs to grain the state and time spaces completely. This was already understood in Ancient Greek and Chinese science. Recall Zeno's paradoxes [86], and the Eudoxus of Cnidus and Liu Hui method of exhaustion [57, 94]. *Symbolic dynamics* was invented [47, 73] to solve problems of dynamics. It is a tool of nonlinear dynamics to analyze models by partitioning of space. The partitioning is supposed to be finite and the parts are labeled with different symbols. The parts are followed in a specially chosen order to make conclusions for not the entire dynamics, but particular features. As we mentioned above, the first ideas of symbolic dynamics can be credited to ancient mathematicians. Next, let us provide support for these ideas.

We will provide arguments that symbolic dynamics was involved in the mathematical discussion much earlier than one might think. There is no sense in arguing that a huge volume of ancient knowledge has to be lost if we trust only modern notations and terms. This is true also for mathematics. Consequently, we must not ignore the priority of results because of differences in terminology. Thus, in the next example let us interpret in modern language what we know from Archimedes, Eudoxus of Cnidus [57] and Liu Hui [94] as the method of exhaustion.

Example 1.2.1. *(decimal approximations) To be convincing we start with the classical example of symbolic dynamics. Consider the finite alphabet consisting of digits* 0, 1, 2, ... , 9 *and an infinite sequence* $a_0...a_1a_2a_3...$ *of the digits. Denote the sequence as an element* s_0*. The Bernoulli shift maps the element to the sequence* $s_1 = a_1...a_2a_3a_4...$. *Proceeding in this way one can obtain the orbit* s_k, $k = 0, 1, 2, ...$ *of the symbolic dynamics* [100]. *To satisfy the general description of symbolic dynamics, one can say that the orbit of the Bernoulli shift is whirling among ten sets of infinite sequences, and each set contains sequences starting with the same digit.*

Now we will describe the dynamics, which was already known in Ancient Greek mathematics, and which can also be accepted as symbolic in modern terminology, namely the approximation of real numbers. Presenting the sequence above as the number $0.\,a_0a_1a_2a_3\ldots$, *and starting with the number* $t_0 = 0.\,a_0$ *introduce the Zeno shift newly such that* $t_1 = Z(t_0) = 0.\,a_0a_1$. *Then* $t_2 = Z(t_1) = 0.\,a_0a_1a_2$, *etc. It is clear that we are dealing with the dynamics of the number approximations. There is no doubt that it is a symbolic dynamics.*

Since real numbers are the basis of mathematics, the last example demonstrates that a symbolic dynamics is the result of fundamental mathematical formalization. Moreover, the dynamics of the Zeno shift is present in many physical characteristics of the real world, in particular in those with probability [33]. Thus, what we are suggesting in our research, the *abstract similarity dynamics*, is much more powerful and universal than the symbolic dynamics, first of all, as a basic instrument for sophisticated dynamics.

Abstract similarity dynamics: a description. We start with fundamental definitions and theorems. Let the set $\mathcal{F}$ be given, which consists of elements $\mathcal{F}_{i_1i_2\ldots i_n\ldots}$, such that

$$\mathcal{F} = \{\mathcal{F}_{i_1i_2\ldots i_n\ldots}\colon i_k = 1, 2, \ldots, m,\ k = 1, 2, \ldots\}, \tag{1.1}$$

where m is a natural number and the elements are called labels. In what follows, we call the set $\mathcal{F}$ the *abstract similarity set* or the *label set*. The following sets,

$$\mathcal{F}_{i_1i_2\ldots i_n} = \bigcup_{j_k=1,2,\ldots,m} \mathcal{F}_{i_1i_2\ldots i_nj_1j_2\ldots}, \tag{1.2}$$

which are called *rectangles*, will be needed.

It is clear that

$$\mathcal{F} \supseteq \mathcal{F}_{i_1} \supseteq \mathcal{F}_{i_1i_2} \supseteq \cdots \supseteq \mathcal{F}_{i_1i_2\ldots i_n} \supseteq \mathcal{F}_{i_1i_2\ldots i_ni_{n+1}}\ldots,\quad i_k = 1, 2, \ldots, m,\ k = 1, 2, \ldots.$$

That is, the rectangles form a nested sequence. Let us introduce the map $\varphi\colon \mathcal{F} \to \mathcal{F}$ such that

$$\varphi(\mathcal{F}_{i_1i_2\ldots i_n\ldots}) = \mathcal{F}_{i_2i_3\ldots i_n\ldots}. \tag{1.3}$$

Considering iterations of the map, one can verify that

$$\varphi^n(\mathcal{F}_{i_1i_2\ldots i_n}) = \mathcal{F} \tag{1.4}$$

for arbitrary natural number n and $i_k = 1, 2, \ldots, m,\ k = 1, 2, \ldots, n$. Due to the relations (1.3) and (1.4), in what follows we shall say that φ is the *abstract similarity map*. An example of the map φ is known as the Bernoulli shift [100] for the symbolic dynamics. We shall call the triple $(\mathcal{F}, \phi, d)$, where $(\mathcal{F}, d)$ is a compact metric space, the *abstract similarity dynamics*.

We will say that for the rectangles $\mathcal{F}_{i_1i_2\ldots i_n}$ the *diameter condition* is valid if

$$\sup_{i_k=1,\,2,\,\ldots,\,m} \operatorname{diam} \mathcal{F}_{i_1i_2\ldots i_n} \to 0 \text{ as } n \to \infty, \tag{1.5}$$

where $\operatorname{diam} A = \sup\{d(\mathbf{x}, \mathbf{y})\colon \mathbf{x}, \mathbf{y} \in A\}$ for a set A in $\mathcal{F}$.

Consider the function $d(A, B) = \inf\{d(\mathbf{x}, \mathbf{y}): \mathbf{x} \in A, \mathbf{y} \in B\}$ of two nonempty sets A and B in $\mathcal{F}$. The abstract similarity set $\mathcal{F}$ satisfies the *separation condition* of degree n, if there exists a positive number ε_0 such that for arbitrary indices $i_1 i_2 \ldots i_n$ one can find indices $j_1 j_2 \ldots j_n$ such that

$$d\left(\mathcal{F}_{i_1 i_2 \ldots i_n}, \mathcal{F}_{j_1 j_2 \ldots j_n}\right) \geqslant \varepsilon_0. \tag{1.6}$$

Let us give the definition of the unpredictability in the sense of the dynamics $(\mathcal{F}, \varphi, d)$.

Definition 1.2.1. [8] *A point $\mathcal{F}_{i_1 i_2 \ldots}$ from the set $\mathcal{F}$ is unpredictable if there exist a positive number ε_0 and sequences t_m, s_m, of natural numbers both of which diverge to infinity such that $\varphi^{t_m}(\mathcal{F}_{i_1 i_2 \ldots i_k \ldots}) \to \mathcal{F}_{i_1 i_2 \ldots i_k \ldots}$ as $m \to \infty$ and $d(\varphi^{t_m + s_m}(\mathcal{F}_{i_1 i_2 \ldots i_k \ldots}), \varphi^{t_m}(\mathcal{F}_{i_1 i_2 \ldots i_k \ldots})) \geqslant \varepsilon_0$ for each natural number m.*

Theorem 1.2.1. [4] *If the diameter and separation conditions hold, then the similarity map possesses Poincarè chaos.*

A point $\mathcal{F}_{i_1 i_2 i_3 \ldots}$ from $\mathcal{F}$ is periodic with period n if its index consists of endless repetitions of a block of n terms. The following theorems are valid.

Theorem 1.2.2. [4] *If the diameter and separation conditions hold, then the abstract similarity map is chaotic in the sense of Devaney.*

If the diameter and separation conditions are valid, then the dynamics $(\mathcal{F}, \varphi, d)$ is said to be *abstract similarity chaotic*. In addition to the Devaney and Poincaré chaos, it can be shown that the Li–Yorke chaos is also present in the dynamics of the map φ. Thus saying that abstract similarity chaos is present means that the three basic types of chaos, Poincaré, Li–Yorke and Devaney chaos, are valid for the dynamics.

How to apply abstract similarity dynamics: domain structured dynamics. The method of abstract similarity dynamics promises to become a general tool of theoretical analysis and applications. It is another, higher, level of abstraction for sophisticated dynamics research. For the moment we can see the three types of problems to which the method can be applied. They depend on how the abstract similarity maps relate to sets, where the dynamics occurs.

Assume that (F, d) is a fixed metric space and consider various relations of the space with an abstract similarity dynamics:

- There is no motion given for the space and one can insert abstract similarity dynamics in the space through the labeling.
- For a dynamical system (F, ψ, d), there exists a label set $\mathcal{F}$ such that the abstract dynamics $(\mathcal{F}, \varphi, d)$ mirrors the motions in F.
- For a random dynamics with a state space F one can find an abstract similarity dynamics, the trajectories of which mirror all realizations of the stochastic process or some of them.

Consider the three modes in more detail. The first one is when a metric space (F, d) is not given a dynamics. Then, by considering the abstract similarity dynamics $(\mathcal{F}, \varphi, d)$ chosen properly for the metric space, we establish a desired behavior in the space (*structural dynamization*), for example, chaos. In our papers this has been done for fractals. We believe that the results are of significant interest, since it is demonstrated that fractals can be domains of chaos in general. Moreover, it is shown in our research that chaos can be arranged in finite dimensional cubes and consequently for any set which is topologically equivalent to the cubes. Another surprise is that chaos can be formulated for finite sets and even two-point sets. This creates nice opportunities to look for chaos ingredients in stochastic processes. Generally speaking the approach is a fundamental contribution to recognize chaos as the most generic type of motion in the world. This is very important for seeking chaotic dynamics in neural networks, since it is needed for the effectiveness of brain activity and artificial intelligence.

Another type is domain structured dynamics applications for models which already are equipped with mappings or vector fields. That is, when a dynamics (F, ψ, d) under analysis can be formalized with an abstract similarity dynamics $(\mathcal{F}, \varphi, d)$ such that the abstract dynamics reflects certain features of the original motion and we can make decision on their existence (*chaos verification*). In this introduction the presentation means that each element of the set F is labeled through a single element of the set $\mathcal{F}$. Generally, we do not assume that uniqueness is required for the relation. Moreover, we keep the distance d for the set $\mathcal{F}$ considering $d(\mathcal{F}_{i_1 i_2 \ldots i_n \ldots}, \mathcal{F}_{j_1 j_2 \ldots jn, \ldots}) = d(f_1, f_2)$, if $\mathcal{F}_{i_1 i_2 \ldots i_n \ldots}$ and $\mathcal{F}_{j_1 j_2 \ldots jn, \ldots}$ are labels of elements f_1 and f_2 of the set F, respectively. This domain structured chaos relates to the models in three modes. The first one relates to models which have not been proven to have any chaos and the domain structured dynamics method is to be tried. Another set is of models which are shown to have one of the conventional types of chaos, but not all, and the abstract similarity chaos confirms presence of all others. Finally, we can consider models such that the domain structured dynamics just confirms what is already known through conservative methods, but does this more effectively and easily.

Let us provide the algorithm which is at the basis of the confirmation. Assume that the metric space (F, d) is under a map ψ such that a dynamical system (F, ψ, d) is obtained. In addition to this consider a label set $\mathcal{F}$ for the set F and the abstract similarity dynamics $(\mathcal{F}, \varphi, d)$ accordingly. Here, we require that the maps ψ and φ *agree*. That is, the value of map ψ at an element of F and the value of similarity map φ at the label of the element coincide. Then, it is clear that certain properties of the motions are the same due to isometry. This is, in particular, true about chaotic features. The dynamics assigned for the triple (F, ψ, d) by the abstract similarity chaos of $(\mathcal{F}, \varphi, d)$ is said to be the *domain structured chaos*. The simplicity, which is in background of our suggestions, makes it possible to consider the features in random processes as well (stochastic processes with chaotic features). The only question is how to find an appropriate labeling. In this way we also consider stochastic processes with continuous and discrete time. All the possibilities are illustrated in this book. In this introductory chapter we will illustrate the theory with the simplest examples.

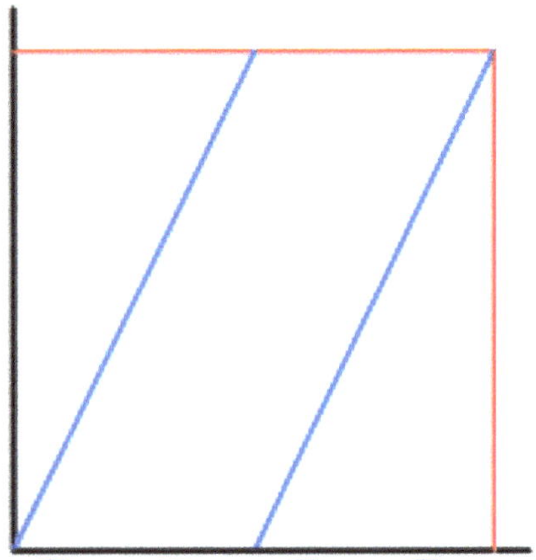

Figure 1.2. The graph of the function B.

Example 1.2.2. *(two-fold scalar baker transformation) Consider the baker map, which is equal to* $B(x) = 2x$ *on* $(0, 1/2]$ *and* $B(x) = 2x - 1$ *on* $(1/2, 1]$*. The graph of the map is seen in figure* 1.2*. For simplicity, we start with immediate construction of the abstract rectangles. Denote* $\mathcal{F}_0 = (0, 1/2]$ *and* $\mathcal{F}_1 = (1/2, 1]$*. Next denote* $\mathcal{F}_j = \mathcal{F}_{j0} \cup \mathcal{F}_{j1}, j = 0, 1,$ *where* $B(\mathcal{F}_{jk}) = \mathcal{F}_j, k = 0, 1$*. Continue, considering inductively* $\mathcal{F}_{i_1 i_2 \ldots i_{n-1} i_n} = \mathcal{F}_{i_1 i_2 \ldots i_n 0} \cup \mathcal{F}_{i_1 i_2 \ldots i_n 1}$*, such that* $B(\mathcal{F}_{i_1 i_2 \ldots i_n k}) = \mathcal{F}_{i_1 i_2 \ldots i_{n-1} k}, k = 0, 1,$ *if all sets* $\mathcal{F}_{i_1 i_2 \ldots i_{n-1} i_n}$ *for a fixed n are determined. The label set* $\mathcal{F}$ *is the collection of points* $\mathcal{F}_{i_1 i_2 \ldots}$ *such that a point with fixed* $i_1 i_2 \ldots$ *belongs to* $\mathcal{F}_{i_1 i_2 \ldots i_n}$ *for all natural n. The reader can easily verify that the set is a chaotic structure. The separation condition is valid for sets* $\mathcal{F}_{ij}, i, j = 0, 1,$ *with* $\epsilon_0 = 1/4$*. This is seen in the figure. A final and the essential remark is that the map B and the similarity map* φ *agree at each point of the domain such that* $B(x) = \varphi(\mathcal{F}_{i_1 i_2 \ldots})$ *if the number* $x \in [0, 1]$ *is labeled with* $\mathcal{F}_{i_1 i_2 \ldots}$*. Thus, the domain structured chaos is proved for the baker transformation. It is notable that the domain of chaos is the unit section and is not a fractal set.*

In chapter 3 we will consider chaos for general multi-fold vector baker transformations.

Example 1.2.3. *Next we will prove the domain structured chaos for the logistic map* $L(x) = mx(1 - x)$ *with the coefficient m larger than four. One can see the graph of the map in figure* 1.3*. We shall label with* $\mathcal{F}$ *the set of all positively invariant points of the unit section. That is, those which are initial points for infinite positive semi-trajectories in the unit. For convenience we do not distinguish the points of the Cantor set and their labels in what follows. The part of the set in the left and right halves of the unite section denote* $\mathcal{F}_0$ *and* $\mathcal{F}_1$*, respectively. Then, find that* $\mathcal{F}_j = \mathcal{F}_{j0} \cup \mathcal{F}_{j1}, j = 0, 1,$ *where* $L(\mathcal{F}_{jk}) = \mathcal{F}_j, k = 0, 1$*. Proceed with the process, considering inductively* $\mathcal{F}_{i_1 i_2 \ldots i_{n-1} i_n} = \mathcal{F}_{i_1 i_2 \ldots i_n 0} \cup \mathcal{F}_{i_1 i_2 \ldots i_n 1}$*, such that* $L(\mathcal{F}_{i_1 i_2 \ldots i_n k}) = \mathcal{F}_{i_1 i_2 \ldots i_{n-1} k}, k = 0, 1,$ *if all sets*

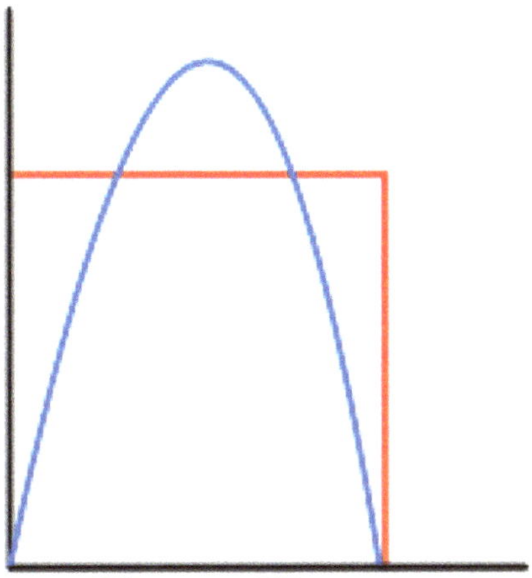

Figure 1.3. The graph of the logistic map.

$\mathcal{F}_{i_1i_2\ldots i_{n-1}i_n}$ *for a fixed n are determined. The label set* $\mathcal{F}$ *is the collection of points* $\mathcal{F}_{i_1i_2\ldots}$. *To prove the diameter property, consider the points* $x_n \to 0$ *as* $n \to \infty$ *such that* $L^n(x_n) = m/4$. *The existence of the sequence proves the property. The separation condition is valid for* $\mathcal{F}_j$, $j = 0, 1$. *Thus, the domain structured chaos is proved for the dynamics.*

Remark 1.2.1. *The last two examples illustrate that the proof of the presence of chaos for conventional models is significantly easier using domain structured dynamics. We suggest also the following open problem: period three implies domain structured chaos.*

Abstract similarity dynamics versus symbolic dynamics. Symbolic dynamics was initiated in the famous paper [73] by Morse and Hedlund. The deep analysis of the obtained structures, in many senses, is a theoretical investigation that repeats of those for continuous dynamics in [31]. The recurrence and transitivity are proved and the recurrence is specified very carefully, as the particular properties of the symbolic dynamics allow this. From one side, the theory of abstract similarity dynamics can be developed like the research of Morse and Hedlund, but with a new, much higher, level of generalization. We hope that the significant research results in [73] will help us to structure the chaos and understand the unpredictable motion more profoundly.

Next, we shall discuss the principal novelty of the abstract similarity dynamics and domain structured dynamics and how they relate to the symbolic dynamics.

The method of conventional abstract symbolic dynamics is beneficial for discrete-time systems with continuous state spaces. The research starts with *finite* partitioning of a state space and each part is assigned its own label. Then, each point of the state space generates an infinite sequence of symbols, the label of the original part to which the start point belongs, the symbol of the part to which the first iteration belongs, and so forth. It is obvious that the original dynamics cannot be followed completely in this way by the symbolic dynamics. Then, why it is useful? One reason is that the dynamics can be drastically simplified. Let us introduce the algorithm of simplification. If x is an element of the state space and $f(x)$ the iteration, then we need a function which takes states to symbols, g, to create the symbol sequence $s = g(x)g(f(x))g(f^2(x))g(f^3(x))\ldots$, that is, $s_0s_1s_2s_3\ldots$. Thus the dynamics is now reduced to shifting in the sequences to the right. If the function g is a homeomorphism or diffeomorphism, then a qualitative result for the symbolic dynamics can be easily reformulated for the original iterations of the map f. Thus, the main difficulty of the method is to find the map and prove that it is a homeomorphism. In our case we do not consider the symbolic sequences at all. We replace the mapping with the procedure of labeling for *all* elements of the state space. One can say that we have extended the partitioning ultimately from the finite to infinite, even *uncountable* partitioning. Accordingly, an abstract similarity map is introduced. Next we ask that the map of the original dynamics agrees with the abstract similarity map. At first glance, the task of such partitioning and the abstract similarity map choice seem difficult, even impossible, but, surprisingly, good results have been achieved in this direction. Moreover, we believe that the class of dynamics the domain of which can be labeled is significantly large. One more important remark is that the total labeling makes the study of dynamics completely reduce to domain structuring, while conservative methods of symbolic dynamics consider only particular properties. To make our method valuable for chaos investigation we have formulated for the set of labels the separation condition and the diameter condition, which guarantee Poincaré, Devaney and Li–Yorke chaos for the dynamics. That is, the domain structured chaos means that the three types of chaos are present and we believe that the method will provide much more interesting results in the future The ultimate labeling makes the map of the state space to the space of sequences of symbols unnecessary and thus the theoretical discussion becomes much more effective.

It is important to clarify what the principal differences of abstract similarity dynamics and symbolic dynamics are and the advantages of the domain structured dynamics, emphasized in the following:

- The state space of symbolic dynamics consists of infinite sequences of symbols. A state space of domain structured dynamics is a set of any nature if its elements can be labeled in the way we have suggested above. We consider also multiple labeling such that even finite sets can be considered as a domain of models. Thus, all sets, finite, countable and uncountable, can be considered for domain structured dynamics, while the symbolic dynamics applications require uncountable domains, since the homeomorphism/diffeomorphism is the equivalence instrument.
- The results of symbolic dynamics are applicable for a dynamics by arranging topological equivalence. The domain structured dynamics is applicable by direct labeling, which is isometery by construction.
- There is the perception that finding and generating partition and one-to-one correspondence between states and the symbol sequences makes research in the symbolic dynamics area completely equivalent to the study of the original

dynamics. This is wrong in the case of a finite partition. The coincidence of all dynamical properties is true for the domain structured dynamics, due to the complete labeling, under the condition that the mappings agree, that is, the labeling of each point of the domain, but not a finite number of its regions. Another benefit of our approach is that the labeling is not necessarily one-to-one. We have applied one-to-several and one-to-infinity labels in our research. This supports essential developments of chaotic ingredients for random processes.

- A Bernoulli shift is directed motion along a sequence of symbols. The abstract similarity map is the motion among elements of a space. There is no special direction. That is, the similarity map is a universal transformation and is restricted only by possibility and choice of labeling. Moreover, for one and the same space set we can apply various label-sets, choosing them as is convenient for a specific problem solution, if needed.
- By considering abstract hyperbolic sets for domain structured dynamics with bi-infinite indexes, we approve chaos for Smale horseshoes and two-dimensional baker transformations in an easier way than conservative methods do. It is not difficult to guess that the abstract hyperbolic structure can be recognized in other dynamics such as the Plykin attractor, Anosov diffeomorphism, etc.
- The domain structured dynamics is an effective tool to find chaotic features in random processes. In our papers and in the current book, laws of large unpredictable strings have been introduced for Bernoulli schemes. Unpredictability is indicated for Bernoulli schemes and Markov chains as well as for processes with continuous time through modular chaos.

The dimension of chaos domains. Let us consider the geometry of domains for different types of chaos. That is, sets of space points where chaotic scenarios occur and which are usually intuitively clear, but theoretically not often so, are specified. To describe the domain for Devaney chaos one can use a trajectory which is dense in the set of all chaotic orbits. Then the closure of the orbit is considered as the domain for the chaos. Otherwise, unite all points which are initial for the sensitivity property, then add the trajectory, which is dense in the set to construct the domain. One can also use the third description of the domain as the union of all unstable cycles of the chaos and take closure of the set as the domain for the Devaney chaos. Similarly, the domain of Li–Yorke chaos consists of the scrambled set united with all cycles of the chaos. It is useful to remember that for the conservative chaos domains can be fractals, that is with a fractional dimension and a dimension which is less than the dimension of equations and larger than the dimension of the trajectory. Possibly the most simple description of the domain is for the Poincaré chaos. It is the quasi-minimal set, that is, the closure of an unpredictable trajectory. The domain structured chaos implies Poincaré, Devaney and Li–Yorke chaos. Hence, all the descriptions above are suitable to describe the domain of the domain structured chaos. Most importantly, the domain structured chaos is in the privileged state, since we determine a domain for the domain structured chaos as the primary step in the analysis. That is, we accept as a domain for the chaos a set which can be labeled in this way. Next we check if the basic diameter and separation conditions are valid. Thus, we have obtained the exceptionally simple algorithm of the chaos domain construction. This implies that we are in circumstances, where profound research can be performed and unexpected results can be obtained. The domain structured chaos can be arranged in any finite dimensional cube from one side and for a domain which consists of two points. That is, the dimension of the chaos domain can be integer and equal to the dimension of the dynamics where the chaos occur. Let us recall that the cube is not fractal, while fractals are considered as very common domains for many chaotic models. The dimension of a fractal is not an integer. The fractals emerge due to the chaos ingredients and continuous maps. The abstract similarity map provides the possibility to consider the domain structured chaos in stochastic processes. We claim that one can organize the chaotic dynamics in any domain if it is of not less than two distanced points.

1.3 Random processes

We are looking for necessary signs of chaos in stochastic processes such as Bernoulli schemes, Wiener or Brownian motion processes, the Poisson processes and Markov chains. It is not surprising that the two antagonistic types of irregular motions, deterministic chaos and stochastic processes, are together subjected to detailed research in many papers. There are different tasks in such investigations. The first is to find common statistical characteristics for the dynamics which are originally different. Another is to randomize deterministic chaos and determine what kind of features are invariant under the perturbations. The third is to find isomorphic dynamics. It does not matter if they are deterministic or stochastic, there exists isomorphism, which makes them abstractly the same.

Our approach to the relations between stochastic processes and deterministic chaos can be understood easily if one remembers that the theory of probability is seeking necessary features in dynamics with chance. In other words, one has to agree that there are no purely deterministic and random processes. The ideal necessary and chance dynamics occur only theoretically. Moreover, necessary elements are even present in theoretical analysis for processes with chance. Accordingly, in our research it has been proved that the ingredients of chaos exist necessarily in stochastic processes such as Bernoulli schemes or Markov chains. In the research the main role is given to the unpredictability. The unpredictable motion is privileged, since its definition absorbs individually all the features needed for chaos. Consequently, one can make a decision on the chaotic nature of an isolated realization of stochastic dynamics. One must specify that when we speak about chaos in the stochastic dynamics, we mean the ingredients which are used for the description of chaos. That is, we prove that the ingredients appear necessarily in processes with chance. For example, the appearance of a finite realization is a necessary event for a random process and the next level of necessity is the occurrence of the unpredictability for a certain random dynamics such as the Bernoulli scheme.

We have already obtained several results in this way. One of the most significant is the concept of unpredictable strings and the two laws of unpredictable strings, which were introduced and proved in [22]. The second law is analogous to the Bernoulli theorem. Next, we find that any finite realization of the Bernoulli scheme as well as of the Markov chain with finite state space is an arc of the unpredictable trajectory of the corresponding deterministic dynamics. The method of seeking the ingredients of chaos was extended for continuous time random processes. Randomly determined unpredictable functions are an important contribution to the theory of unpredictable functions. Thus, the class of functions which is needed for chaos research and applications has been significantly enlarged. Generally speaking, we have moved the research to the area of the theory of probability in many senses. Finally, the construction of unpredictable functions as realizations of random processes will create a new line of research for stochastic differential equations. The first steps in the research have already been taken in our papers. We investigate the existence of unpredictable solutions of equations, for which perturbations are unpredictable functions generated randomly. Another interesting problem to study in the future is the statistical characteristics for concrete random processes with discrete and continuous time on the basis of deterministic parameters provided by chaos investigation.

1.4 Abstract fractals

Fractals and chaos are twins, which are growing in different families, geometry and dynamics. The domains of chaos are sets, that is collections of points, where the dynamics occurs, and they are in many cases fractals. That is, the domains for difference or differential equations are either self-similar sets or of fractional dimension. The famous example is the Cantor set, the domain for the logistic map. Intuitively, it is clear that the porosity is a consequence of sensitivity and continuity in the initial state. Is any fractal suitable as a domain for a chaotic motion? A positive answer is provided in our papers, where domain structured chaos was built on such fractals as the Sierpinski carpet and triangle, the Koch snowflake and others. However, it is curious that the domain structured chaos can be constructed in finite dimensional Euclidean cubes, which are not fractals.

The abstract similarity brings chaos and fractals closer. This is why we start with the history of the concept.

It is convention to say that several objects, abstract or real, are *similar* to each other if they share common properties. Mathematical similarity is a relation between two or several point sets. For example, two geometric figures are similar if there exists a similarity transformation that maps one onto the other [96]. If the similarity is considered within a single object, it is called self-similarity. Geometrically the self-similarity is understood as a property of figures for which parts are similar to the whole at all scales [28, 55]. The origins of self-similarity date back to the seventeenth century, when Leibniz introduced the notion of recursive self-similarity [16]. The nineteenth century witnessed a lot of work on the subject, embodied in several discoveries such as the Weierstrass function by Karl Weierstrass, the Cantor set by Georg Cantor and the space-filling curves described by Giuseppe Peano and David Hilbert. Other examples of self-similar sets are the Koch curve, discovered by Helge von Koch in 1904, the Sierpinski gasket and carpet, which were introduced by Waclaw Sierpinski in 1916 and the Lévy C curve described by Paul Lévy in 1938. Benoit Mandelbrot invented the term '*fractal*' in 1975 [70, 71] to describe certain geometrical structures that exhibit self-similarity. Since then this term has been employed to denote all the above mentioned sets and the field became known as *fractal geometry*. Consequently, the fractal concept is axiomatically linked to the notion of self-similarity, which is considered to be one of the acceptable definitions of fractals. That is, a fractal can be defined as a set that displays self-similarity at all scales. However, Mandelbrot defines a fractal as a set whose Hausdorff dimension is strictly larger than its topological dimension [48, 71, 72]. In summary, self-similarity and the fractional dimension are the two most important features of fractals. The connection between them is that self-similarity is the easiest way to construct a set that has a fractional dimension [38].

The idea of a *self-similar set* was first considered by Moran in 1946 [77], who gave a mathematical definition of a geometric construction as a collection of sets satisfying specific conditions. Moran's construction has been studied extensively and developed by many researchers proposing various methods. In [82, 83, 92] the approach was generalized to build different classes of Moran-like geometric constructions with the help of a symbolic space. The topological conjugacy to a symbolic dynamical system was employed to describe the symbolic representation of the constructions. This type of self-similar set is referred to as the Cantor-like set. Interesting definitions of self-similarity and related problems of dimension and measure are discussed in the papers [27, 41, 43, 44, 49, 54, 59, 80, 91]. These publications consider sets in Euclidean space $\mathbb{R}^n$ and define a self-similar set as a union of its images under similarity transfunctions [44].

Fractal geometry is a mathematical tool used to describe many natural structures that adopt various degrees of *self-similarity*, as well as to design some artificial structures. A set that displays self-similarity and repeats the same patterns at every scale is usually called fractal. Mandelbrot defined a fractal as a set for which the Hausdorff dimension strictly exceeds the topological dimension [71]. Dealing with fractals goes back to the seventeenth century when Leibniz introduced the notions of recursive self-similarity [103]. Since then, history did not record anything about self-similarity until the late nineteenth century when Weierstrass introduced in 1872 a function that was everywhere continuous but nowhere differentiable. The graph of the Weierstrass function became an example of a self-similar fractal. The Cantor set, constructed by Cantor in 1883, is considered as the most essential and influential fractal, since it is a simple and perfect example for the theory and applications of fractals. The idea of self-similarity received more attention in the work of von Koch. He devised in 1904 a continuous but non-differentiable curve that never intersects itself. The curve is considered as one of the simplest regular fractals. Sierpinski was one of the mathematicians who made significant contributions in the field of fractals. He introduced the famous triangular fractal in 1916, known as the Sierpinski gasket. The fractal is generated by a recursive process of removing symmetrical parts from an initial triangle. In an analogous way to the gasket, Sierpinski developed a square fractal known as the Sierpinski carpet. Julia sets gained significance in being generated using the dynamics of an iterative function. They were discovered by Gaston Julia and Pierre Fatou in 1917–19, where they studied independently the iteration of rational functions in the complex plane [72]. During the same period the mathematician Felix Hausdorff formulated the notion of fractional dimension which became a very important tool for studying and characterizing the geometrical complexity of fractals. Paul Lévy studied self-similar curves and surfaces and in 1938 he described a new fractal curve, the Lévy C curve.

Several researchers pointed out that a close relationship between chaos and fractal geometry can be observed. This can be seen, for example, in the dynamics of the Fatou–Julia iteration used to construct Julia and Mandelbrot sets where two neighboring points in the domain which are close to the boundary may have completely different behavior. That is, we can speak of sensitivity in fractal structures. Chaos tells us about the state of irregularity and the divergence of trajectories which depend on the nature of the dynamics, whereas the fractal concept can be used to study complex geometric structures. Therefore, the link between chaos and fractals is more clear when the fractal dimension is used to measure the extent to which a trajectory fills its phase space. In other words, the fractal dimension of the orbit in phase space implies the existence of a strange attractor [76]. The fundamental work on the chaotic nature of fractals has been done only for specific types categorized under the totally disconnected fractals [29, 36, 39]. In that work, the topological conjugacy concept was utilized to prove that these fractal sets are invariant for certain chaotic maps. Apart from that, relatively few studies have been carried out on chaotic dynamical systems for fractals and perhaps the most relevant one is the work on the Sierpinski carpet in [42]. In that research the author shows that the dynamical system associated with a shift transformation defined on the Sierpinski carpet set is chaotic in the sense of topological mixing.

1.5 Differential equations

The following series of implications is the basis for the integrity of this book:

Abstract similarity dynamics → *Domain structured chaos* → *Random processes* → *Unpredictable functions* → *Unpredictable solutions of differential equations* → *Unpredictable oscillations in neural networks* → *Poincaré chaos in neural networks.*

Observing the line of implications, one can conclude that by reading this book in sequence, we pass from very abstract constructions to the analysis of the most essential real world problems. The last two chapters of this book provide strong evidence that domain structured chaos is useful for applied mathematics. We have demonstrated through the present research that randomness is potentially a rich source of unpredictable functions and that the abstract similarity dynamics is a tool which can be utilized for theoretical support of the constructions. The process of the interaction of the abstract similarity dynamics and stochastic processes is two-fold. Not only can stochastic processes serve to obtain unpredictable functions, but unpredictable functions are useful to build structures for random dynamics. Thus, new methods of analysis for stochastic differential equations can appear. Now we shall discuss not stochastic, but deterministic differential equations. It is usual that oscillations are the main subject for theories of differential equations. This is caused, first of all, by application demands. This is why, theoretically, it is one of the most important duties of mathematicians to find analytical presentations of the oscillations and as many as possible. Unfortunately, this task is not easy for motions with weak recurrency. It is exceptionally difficult for Poisson stable functions. Hence, one could not find developed samples of the functions, presented analytically, or even numerical algorithms for simulation. We have supported our research with graphical illustration of unpredictable functions as solutions of differential equations, asymptotically. This is proved rigorously in examples 6.4.1, 6.4.2 and 2.2.1. Thus one can say that not only unpredictable functions but also Poisson stable functions are exemplified and simulated in our research for the first time. These achievements are based on our experience in the study of hybrid systems or, more precisely, on differential equations with various types of discontinuity. The issue points to the construction of unpredictable functions through the method of domain structured chaos. It is seen in examples 2.1.5, 2.2.1, 1.2.3, 3.1.5 and 3.3.2. We are strongly confident that the supply of unpredictable functions will be enlarged significantly by utilizing functions determined randomly on the basis of the abstract similarity dynamics and verification of the chaotic ingredients in random processes. This will make theoretical research of unpredictable solutions for dynamical equations very valuable for applications, first of all in modern research areas such as computer sciences, artificial networks, neural networks, brain activity and others. Mathematically, we predict that the study of unpredictable solutions will be extended by methods of operator theory, averaging methods, singular differential equations and numerical methods. Consequently, the results can be easily exported for partial differential equations, functional–differential equations, evolution equations and various types of hybrid systems.

It is worth noting that in the literature a large number of results are obtained for periodic, quasi-periodic and almost periodic solutions of differential equations using the established mathematical methods and important applications [32, 37, 45, 46, 48]. On the other hand, recurrent and Poisson stable solutions are also crucial for the theory of differential equations [35, 88, 90]. Our proposal can revive the interest of specialists in differential equation theory for two reasons. The first is related to the verification of the unpredictability which requires a more sophisticated technique than for recurrent and Poisson stable solutions. Thus the problem of the existence of unpredictable solutions is a challenging one. Unpredictable solutions can be investigated for various types of differential equations such as partial differential equations, evolution equations and hybrid systems. That is, a new approach for chaos extension into many types of dynamics is suggested in our study. In [90] a general method of the comparability of functions through the character of their recurrence was suggested for differential equations. In particular, it is popular for solutions with Poisson stability. In this book, and in papers [10–13], we apply the approach to prove the existence of Poisson stable solutions, which is much more straightforward, and consequently effective, than that in [35, 90]. It can be applied in the future to various types of dynamical equations.

1.6 Neural networks

Why is a human mind able to reflect the world around it and be creative to make significant changes in the world? The most popular explanation says that the large number, one hundred billion, of neurons is the reason for the mightiness of the mind. Another explanation says that the mental power is created by the numerous connections of synapses, numbering several trillions. We believe that the reason is chaotic motions, which occur in the electrical activity of neural networks, since a chaos of networks with a finite number of neurons contains infinitely many cycles as orbits. If a single cycle is already a portion of information, one can imagine how large is the volume of information in a single neural network with infinite periodic motions, which can be uncountable. In our papers we show that not only fixed points or cycles as recurrent motions are present in chaotic dynamics. First, we have demonstrated that infinitely many dense almost periodic motions can form the basis for a chaotic dynamics. Finally, we strictly proved that an uncountable set of unpredictable oscillations can be at the basis of chaos.

Oscillations are ubiquitous in brain activity and they strongly correlate with many cognitive tasks. For example, neural networks in the mammalian forebrain demonstrate several oscillatory bands covering frequencies from approximately 0.05 Hz to 500 Hz. Detailed biophysical investigations have revealed that even single neurons demonstrate the ability to resonate and oscillate, thus one can suggest that the activity within neural networks represents information [53, 67]. Oscillatory neural networks are effective for image recognition [85] and in activating network states associated with memory recall [87]. It is natural that neural oscillations have become the core of the interdisciplinary research that unites neuroscience, psychophysics, biophysics, cognitive psychology, computational modeling, mathematics and physics [25, 26, 30, 34, 58, 60, 69, 75, 93, 95, 98, 101]. When studying exogenous oscillations, one must talk about oscillatory neural networks [50, 52, 61, 74, 85, 89]. Moreover, considering the role of fluctuations in neural network activity, one must focus on the endogenous oscillations. This is why many papers are devoted to periodic, almost periodic and unpredictable oscillations, and exponential stability for neural networks considering the input/output mechanisms [2, 5–7, 17, 18, 21, 23, 24, 40, 51, 56, 62–66, 81, 97, 99, 102].

References

[1] Addabbo T, Fort A, Rocchi S and Vignoli V 2011 Digitized chaos for pseudo-random number generation in cryptography *Chaos-based Cryptography: Theory, Algorithms and Applications* ed L Kocarev and S Lian (Berlin: Springer) pp 67–97

[2] Aihara K, Takabe T and Toyoda M 1990 Chaotic neural networks *Phys. Lett.* A **144** 333–40

[3] Akhmet M and Alejaily E M 2019 Domain-structured chaos in a Hopfield neural network *Int. J. Bifurcat. Chaos* **29** 1950205

[4] Akhmet M and Alejaily E M 2021 Abstract similarity, chaos and fractals *Discrete Contin. Dyn. Syst.* B **26** 2479–97

[5] Akhmet M, Aruğaslan D and Yilmaz E 2010 Stability analysis of recurrent neural networks with piecewise constant argument of generalized type *Neural Netw.* **23** 805–11

[6] Akhmet M, Aruğaslan Çinçin D, Tleubergenova M and Nugayeva Z 2021 Unpredictable oscillations for Hopfield-type neural networks with delayed and advanced arguments *Mathematics* **9** 1–19

[7] Akhmet M and Fen M O 2016 *Replication of Chaos in Neural Networks, Economics and Physics* (Berlin: Springer)

[8] Akhmet M and Fen M O 2016 Poincaré chaos and unpredictable functions *Commun. Nonlinear Sci. Numer. Simul.* **48** 85–94

[9] Akhmet M and Fen M O 2016 Unpredictable points and chaos *Commun. Nonlinear Sci. Numer. Simul.* **40** 1–5

[10] Akhmet M and Fen M O 2017 Existence of unpredictable solutions and chaos *Turk. J. Math.* **41** 254–66

[11] Akhmet M and Fen M O 2018 Non-autonomous equations with unpredictable solutions *Commun. Nonlinear Sci. Numer. Simul.* **59** 657–70

[12] Akhmet M, Fen M O and Alejaily E M 2020 *Dynamics with Chaos and Fractals* (Berlin: Springer)

[13] Akhmet M, Fen M O, Tleubergenova M and Zhamanshin A 2019 Unpredictable solutions of linear differential and discrete equations *Turk. J. Math.* **43** 2377–89

[14] Akhmet M, Fen M O, Tleubergenova M and Nugayeva Z 2020 Unpredictable solutions of linear impulsive systems *Mathematics* **8** 1798

[15] Akhmet M, Fen M O, Tleubergenova M and Zhamanshin A 2019 Poincaré chaos for a hyperbolic quasilinear system *Miskolc Math. Notes* **20** 33–44

[16] Zmeskal O, Dzik P and Vesely M 2013 Entropy of fractal systems *Comput. Math. Appl.* **66** 135–46

[17] Akhmet M, Seilova R, Tleubergenova M and Zhamanshin A 2020 Shunting inhibitory cellular neural networks with strongly unpredictable oscillations *Commun. Nonlinear Sci. Numer. Simul.* **89** 105287

[18] Akhmet M, Tleubergenova M and Nugayeva Z 2020 Strongly unpredictable oscillations of Hopfield-type neural networks *Mathematics* **8** 1791

[19] Akhmet M, Tleubergenova M and Zhamanshin A Quasilinear differential equations with strongly unpredictable solutions *Carpathian J. Math.* **36** 341–9

[20] Akhmet M, Tleubergenova M and Zhamanshin A 2019 Neural networks with Poincaré chaos *ACM Int. Conf. Proc. Series* 34–7

[21] Akhmet M, Tleubergenova M and Zhamanshin A 2020 Inertial neural networks with unpredictable oscillations *Mathematics* **8** 1797

[22] Akhmet M and Tola A 2020 Unpredictable strings *Kazakh Math. J.* **20** 16–22

[23] Akhmet M and Yilmaz E 2013 *Neural Networks with Discontinuous/Impact Activations* (New York: Springer)

[24] Chen A and Cao J 2002 Almost periodic solution of shunting inhibitory CNNs with delays *Phys. Lett.* A **298** 161–70

[25] Arbib M A 1988 *Brain Theory and Neural Networks* (Cambridge, MA: MIT Press)

[26] Baldi P and Brunak S 2001 *Bioinformatics: The Machine Learning Approach* (Cambridge, MA: MIT Press)

[27] Bandt C and Graf S 1992 Self-similar sets 7. A characterization of self-similar fractals with positive Hausdorff measure *Proc. Am. Math. Soc.* **114** 995–1001

[28] Barabasi L 1995 *Fractal Concepts in Surface Growth* (Cambridge: Cambridge University Press)

[29] Barnsley M 1988 *Fractals Everywhere* (New York: Academic)
[30] Begg R, Kamruzzaman J and Sarker R 2006 *Neural Networks in Healthcare: Potential and Challenges* (London: Idea Group Publishing)
[31] Birkhoff G D 1991 *Dynamical Systems* (Providence, RI: Colloquium)
[32] Bohr H A 1947 *Almost Periodic Functions* (New York: Chelsea)
[33] Bunimovich L and Sinai Y 1980 Markov partitions for dispersed billiards *Commun. Math. Phys.* **78** 247–80
[34] Buzsaki G 1989 Two-stage model of memory trace formation: a role for 'noisy' brain states *Neuroscience* **31** 551–70
[35] Cheban D and Liu Z 2020 Periodic, quasi-periodic, almost periodic, almost automorphic, Birkhoff recurrent and Poisson stable solutions for stochastic differential equations *J. Differ. Equ.* **269** 3652–85
[36] Chen G and Huang Y 2011 *Chaotic Maps: Dynamics, Fractals and Rapid Fluctuations (Synthesis Lectures on Mathematics and Statistics)* (San Rafael, CA: Morgan and Claypool)
[37] Corduneanu C 2009 *Almost Periodic Oscillations and Waves* (New York: Springer)
[38] Crownover R M 1995 *Introduction to Fractals and Chaos* (Boston, MA: Jones and Bartlett)
[39] Devaney R L 1989 *An Introduction to Chaotic Dynamical Systems* (Canada; USA: Addison-Wesley)
[40] Dong Q, Matsui K and Huang X 2002 Existence and stability of periodic solutions for Hopfield neural network equations with periodic input *Nonlinear Anal.* **49** 471–9
[41] Edgar G A 1990 *Measure, Topology, and Fractal Geometry* (New York: Springer)
[42] Ercai C 1997 Chaos for the Sierpinski carpet *J. Stat. Phys.* **88** 979–84
[43] Falconer K J 1985 *The Geometry of Fractal Sets* (Cambridge: Cambridge University Press)
[44] Falconer K J 1995 Sub-self-similar sets *Trans. Am. Math. Soc.* **347** 3121–9
[45] Farkas M 1994 *Periodic Motion* (New York: Springer)
[46] Fink A M 1974 *Almost Periodic Differential Equations* (New York: Springer)
[47] Hadamard J 1898 Les surfaces à courbures opposées et leurs lignes géodésiques *J. Math. Pures Appl.* **5** 27–73
[48] Hartman P 2002 *Ordinary Differential Equations* (New York: SIAM)
[49] Hata M 1985 On the structure of self-similar sets *Japan J. Appl. Math.* **2** 381–414
[50] Hayashi Y 1994 Oscillatory neural network and learning of continuously transformed patterns *Neural Netw.* **7** 219–31
[51] He Y, Wu M and She J-H 2006 Delay-dependent exponential stability of delayed neural networks with time-varying delay *IEEE Trans. Circuits Syst. II Exp. Briefs* **53** 553–7
[52] Hoppensteadt F C and Izhekevich E M 1999 Oscillatory neurocomputers with dynamic connectivity *Phys. Rev. Lett.* **82** 2983–6
[53] Hutcheon B and Yarom Y 2000 Resonance, oscillation and the intrinsic frequency preferences of neurons *Trends Neurosci.* **23** 216–22
[54] Hutchinson J 1981 Fractals and self-similarity *Indiana Univ. Math. J.* **30** 713–47
[55] Janahmadov A K and Javadov M 2018 *Fractal Approach to Tribology of Elastomers* (Cham: Springer)
[56] Jiang H, Zhang L and Teng Z 2005 Existence and global exponential stability of almost periodic solution for cellular neural networks with variable coefficients and time-varying delays *IEEE Trans. Neural Netw.* **16** 1340–51
[57] Knorr W R 1986 *The Ancient Tradition of Geometric Problems* (Basel: Birkhauser)

[58] Koepsell K, Wang X, Hirsch J A and Sommer F T 2010 Exploring the function of neural oscillations in early sensory systems *Front. Neurosci.* **4** 53–61
[59] Lau K S, Ngai S M and Rao H 2001 Iterated function systems with overlaps and the self-similar measures *J. London Math. Soc.* **63** 99–116
[60] Laurent G 2002 Olfactory network dynamics and the coding of multidimensional signals *Nat. Rev. Neurosci.* **3** 884–95
[61] Li R and Zhou X 2013 A note on chaos in product maps *Turk. J. Math.* **37** 665–75
[62] Li X and Song S 2013 Impulsive control for existence, uniqueness, and global stability of periodic solutions of recurrent neural networks with discrete and continuously distributed delays *IEEE Trans. Neural Netw. Learn. Syst.* **24** 868–77
[63] Li Y and Meng X 2018 Almost periodic solutions for quaternion-valued shunting inhibitory cellular neural networks of neutral type with time delays in the leakage term *Int. J. Syst. Sci.* **49** 2490–505
[64] Liao G, Chen X and Sanchez N E 2002 Delay-dependent exponential stability analysis of delayed neural networks: an LMI approach *Neural Netw.* **15** 855–66
[65] Liu B and Huang L 2008 Positive almost periodic solutions for recurrent neural networks *Nonlinear Anal.: Real World Appl.* **9** 830–841
[66] Liu Z and Liao L 2004 Existence and global exponential stability of periodic solution of cellular neural networks with time-varying delays *J. Math. Anal. Appl.* **290** 247–62
[67] Llinas R 1988 The intrinsic electrophysiological properties of mammalian neurons: insights into central nervous system function *Science* **242** 830–41
[68] Lorenz E N 1984 The local structure of chaotic attractor in four dimensions *Physica* D **13** 90–104
[69] Maguire M J and Abel A D 2013 What changes in neural oscillations can reveal about developmental cognitive neuroscience: language development as a case in point *Dev. Cogn. Neurosci.* **6** 125–36
[70] Mandelbrot B B 1975 *Les Objets Fractals: Forme, Hasard, et Dimension* (Paris: Flammarion)
[71] Mandelbrot B B 1983 *The Fractal Geometry of Nature* (New York: Freeman)
[72] Mandelbrot B B 2004 *Fractals and Chaos: The Mandelbrot Set and Beyond* (New York: Springer)
[73] Masoller C, Schifino A C and Romanelli L 1995 Characterization of strange attractors of Lorenz model of general circulation of the atmosphere *Chaos Soliton. Fract.* **6** 357–66
[74] Miller V H and Jansen B H 2008 Oscillatory neural network for pattern recognition: trajectory based classification and supervised learning *Biol. Cybern.* **99** 459–71
[75] Mirollo R E and Strogatz S H 1990 Synchronization of pulse-coupled biological oscillators *SIAM J. Appl. Math.* **50** 1645–62
[76] Moon F C 1992 *Chaotic and Fractal Dynamics: An Introduction for Applied Scientists and Engineers* (New York: Wiley)
[77] Moran P A P 1946 Additive functions of intervals and Hausdorff measure *Proc. Camb. Phil. Soc.* **42** 15–23
[78] Nayfeh A 2000 *Perturbation Methods* (New York: Wiley)
[79] Nemytskii V V and Stepanov V V 1960 *Qualitative Theory of Differential Equations* (Princeton, NJ: Princeton University Press)
[80] Ngai S M and Wang Y 2001 Hausdorff dimension of overlapping self-similar sets *J. London Math. Soc.* **63** 655–72

[81] Ou C 2009 Almost periodic solutions for shunting inhibitory cellular neural networks *Nonlinear Anal.: Real World Appl.* **10** 2652–8
[82] Pesin Y 1997 *Dimension Theory in Dynamical Systems: Contemporary Views and Applications* (Chicago, IL: University of Chicago Press)
[83] Pesin Y and Weiss H 1996 On the dimension of deterministic and random Cantor-like sets, symbolic dynamics, and the Eckmann–Ruelle conjecture *Commun. Math. Phys.* **182** 105–53
[84] Poincaré H 1957 *New Methods of Celestial Mechanics* vol 1–3 (New York: Dover)
[85] Raiko T and Valpola H 2011 Oscillatory neural network for image segmentation with biased competition for attention *Adv. Exp. Med. Biol.* **718** 175–85
[86] Sainsbury R M 2003 *Paradoxes* (Cambridge: Cambridge University Press)
[87] Schmidt H, Avitabile D, Montbrió E and Roxin A 2018 Network mechanisms underlying the role of oscillations in cognitive tasks *PLoS Comput. Biol.* **14** e1006430
[88] Sell G R 1971 *Topological Dynamics and Ordinary Differential Equations* (London: Van Nostrand Reinhold)
[89] Selverston A I and Moulins M 1985 Oscillatory neural networks *Annu. Rev. Physiol.* **47** 29–48
[90] Shcherbakov B A 1969 Poisson stable solutions of differential equations, and topological dynamics *Differ. Uravn.* **5** 2144–55 (in Russian)
[91] Spear D W 1992 Measure and self-similarity *Adv. Math.* **91** 143–57
[92] Stella S 1992 On Hausdorff dimension of recurrent net fractals *Proc. Am. Math. Soc.* 116 389–400
[93] Steriade M 2001 Impact of network activities on neuronal properties in corticothalamic systems *J. Neurophysiol.* **86** 1–39
[94] Straffin P D 1998 Liu Hui and the first golden age of chinese mathematics *Math. Mag.* **71** 163–81
[95] Traub R D, Jeffers J G R and Whittington M A 1999 *Fast Oscillations in Cortical Circuits* (Cambridge, MA: MIT Press)
[96] Volrath H J 1977 The understanding of similarity and shape in classifying tasks *Educ. Stud. Math.* **8** 211–24
[97] Wang L 2010 Existence and global attractivity of almost periodic solutions for delayed high-ordered neural networks *Neurocomputing* **73** 802–8
[98] Ward L M 2003 Synchronous neural oscillations and cognitive processes *Trends Cogn. Sci.* **7** 553–9
[99] Wen S, Huang T, Zeng Z, Chen Y and Li P 2015 Circuit design and exponential stabilization of memristive neural networks *Neural Netw.* **63** 48–56
[100] Wiggins S 1988 *Global Bifurcation and Chaos, Analytical Methods* (New York: Springer)
[101] Winfree A 1980 *The Geometry of Biological Time* (New York: Springer)
[102] Yi Z, Heng P A and Fu A C 1999 Estimate of exponential convergence rate and exponential stability for neural networks *IEEE Trans. Neural Netw.* **10** 1487–93

Part II

Chaos and fractals

IOP Publishing

Domain Structured Dynamics
Unpredictability, chaos, randomness, fractals, differential equations and neural networks
Marat Akhmet

Chapter 2

Unpredictability implies chaos

Sections of this chapter have been reprinted from [2], copyright (2017), with permission from Elsevier.

The first and most impressive evidence of chaos was discovered by Henri Poincaré in celestial mechanics [25]. Surprisingly, he showed that there are no stable cycles in the three-body problem. That is, recurrence, which is less simple than periodic motion, was observed and the ultimate recurrent motion was named after Siméon Poisson [25]. The closure of a compact Poisson stable orbit is the quasi-minimal set [14, 19, 24] and it is the region where very sophisticated dynamics occur. This was the ultimate achievement in the field of recurrence, until the concept of the unpredictability was introduced in [3].

The next significant discovery was made in meteorology, which is far removed from celestial mechanics. American physicist Edward Lorenz [21] learned that the dynamics admits what we know as sensitivity. Unfortunately, sensitivity was not incorporated in quasi-minimal sets for a long period of time, until the *unpredictable point* was introduced in [3]. On one hand it is a specific type of Poisson stable point and on the other hand unpredictability is a new type of recurrence. We think that it is a new ultimate achievement in the field of recurrent motions, which replaced the Poisson stable point in this role since it is endowed with divergence and we consider the unpredictable point as the most mild recurrence which is already behind the gate of chaos.

This chapter is an introductory part of this book. Unpredictability is introduced as a new phenomenon of classical dynamical systems and its appearance in functional spaces is also discussed.

2.1 Unpredictability in dynamics

It is needless to say that the theory of dynamical systems is the research of oscillations, one can also say the research of recurrence, and the most sophisticated motion of the theory is the Poisson stable trajectory [27]. In [3], inspired by the investigation of chaos, we introduced a new type of oscillation, next to the Poisson stable one, and call the

doi:10.1088/978-0-7503-3507-2ch2

initial point for it the *unpredictable point* and the trajectory itself the *unpredictable orbit*. These novelties make possible the connection of homoclinic chaos and the latest types of chaos through the concept of *unpredictability*, that is, sensitivity assigned to a single orbit. Thus the *Poincaré chaos* concept was eventually shaped in our paper [3]. We have also determined the unpredictable function on the real axis as an unpredictable point of the Bebutov dynamics in [2], involving widely differential and discrete equations in the investigation of chaos. However, we need more a precise description, what one should understand as an unpredictable function. The current section is devoted to this constructive task.

There are different types of chaos. The first is homoclinic chaos [29], which is presented in the famous publication of Poincaré [25]. Another is Devaney chaos [15], whose ingredients are transitivity, Lorenz sensitivity [21] and the existence of infinitely many unstable periodic motions which are dense in the chaotic attractor. The third is Li–Yorke chaos [20], which is characterized by the presence of a scrambled set in which any pair of distinct motions are proximal and frequently separated. One can see that the crucial difference between homoclinic chaos and the other types is the sensitivity and frequent separation features. For homoclinic chaos it is assumed that there is a homoclinic structure and, moreover, that there is instability. Lorenz [21] was the first to define the divergence of neighboring motions as sensitivity, a specific sort of instability, and this was followed by Li and Yorke [20] who used frequent separation and proximality for the same purpose. The absence of a quantitative description of instability makes homoclinic chaos different from the later definitions and this causes some inconvenience. This is because sensitivity is assumed to be one of the main ingredients of chaos in its modern comprehension. Poincaré himself was aware of the divergence of initially nearby trajectories, but did not give exact prescriptions as to how it should proceed. This is why one can say that the construction of the puzzle which was initiated and designed by the French genius is still not complete. In our study we are trying to make a contribution to working out this puzzle. For this purpose we have utilized open Poisson stable motions, which accompany homoclinic chaos [29]. In [3] we developed the concept of the Poisson stable point to the concept of the unpredictable point utilizing the *unpredictability* as individual sensitivity for a motion. Thus from a single point of a trajectory we use it as the Ariadne's thread to come to a phenomenon which we call *Poincaré chaos* in [3]. This phenomenon brings all types of chaos closer, since it is another description of motion in dynamics with homoclinic structure and it also admits ingredients similar to the later chaos types, that is, transitivity, sensitivity, frequent separation and proximality. The presence of infinitely many periodic motions in the later definitions can be substituted by a continuum of Poisson stable orbits. Our main hope is that this suggestion may bring the research of chaos back to the theory of classical dynamical systems. The strong argument for this is the fact that we introduced a new type of motion. That is, the already existing list of oscillations in dynamical systems from equilibrium to Poisson stable orbits now includes unpredictable motions. This enlargement of the list will give impetus for the further extension of dynamical systems theory. In applications some properties and/or laws of dynamical systems can be lost or ignored, for example, if one considers non-autonomous or non-smooth systems. Then we can apply unpredictable functions [2], a new type of oscillation which immediately follows almost periodic solutions of differential equations in the rank of bounded solutions. They can be investigated for any type of equation, since through our results they can be treated using the methods of the qualitative theory of differential equations [18].

We utilize the topology of uniform convergence on any compact subset of the real axis to introduce the unpredictable functions. More precisely, the Bebutov dynamical system [27] has been applied. Poisson functions are introduced using the same dynamics [27]. All this makes our task of incorporating the investigation of chaos in the theory of differential equations initiated in papers [2, 5, 7] and in the book [6] proceed in the correct way. The main goal of this chapter is the construction of a concrete unpredictable function. To determine the procedure we start with unpredictable sequences as motions of symbolic dynamics and of the logistic map. Then an unpredictable function is determined as an improper convolution integral with a relay function. Finally, we demonstrate examples of unpredictable functions as solutions of differential equations.

Let (X, d) be a metric space and $\pi\colon \mathbb{T}_+ \times X \to X$, where $\mathbb{T}_+$ is either the set of non-negative real numbers or the set of non-negative integers, be a semi-flow on X, i.e. $\pi(0, x) = x$ for all $x \in X$, $\pi(t, x)$ is continuous in the pair of variables t and x, and $\pi(t_1, \pi(t_2, x)) = \pi(t_1 + t_2, x)$ for all $t_1, t_2 \in \mathbb{T}_+$, $x \in X$.

A point $x \in X$ is called positively Poisson stable (stable P^+) if there exists a sequence $\{t_n\}$ satisfying $t_n \to \infty$ as $n \to \infty$ such that $\lim_{n\to\infty}\pi(t_n, x) = x$ [24]. For a given point $x \in X$ let us denote by Θ_x the closure of the trajectory $T(x) = \{\pi(t, x)\colon\ t \in \mathbb{T}_+\}$, i.e. $\Theta_x = \overline{T(x)}$. The set Θ_x is a quasi-minimal set if the point x is stable P^+ and $T(x)$ is contained in a compact subset of X [24].

It was demonstrated by Hilmy [19] that if the trajectory corresponding to a Poisson stable point x is contained in a compact subset of X and it is neither a rest point nor a cycle, then the quasi-minimal set contains an uncountable set of motions that is everywhere dense and Poisson stable. The following theorem can be proved by adapting the technique given in [19, 24].

Theorem 2.1.1. [3] *Assume that $x \in X$ is stable P^+ and $T(x)$ is contained in a compact subset of X. If Θ_x is neither a rest point nor a cycle, then it contains an uncountable set of motions everywhere dense and stable P^+.*

The definitions of the unpredictable point and unpredictable trajectory are as follows.

Definition 2.1.1. [3] *A point $x \in X$ and the trajectory through it are unpredictable if there exists a positive number ϵ_0 (the unpredictability constant) and sequences $\{t_n\}$ and $\{\tau_n\}$, both of which diverge to infinity, such that $\lim_{n\to\infty} \pi(t_n, x) = x$ and $d(\pi(t_n + \tau_n, x), \pi(\tau_n, x)) \geqslant \epsilon_0$ for each $n \in \mathbb{N}$.*

Markov [24] proved that a trajectory that is stable in both the Poisson and Lyapunov (uniformly) senses must be an almost periodic one. Since definition 2.1.1

implies instability, an unpredictable motion cannot be almost periodic. In particular, it is neither an equilibrium, nor a cycle.

Based on unpredictable points, a new definition of chaos was provided in the [3] as follows.

Definition 2.1.2. [3] *The dynamics on the quasi-minimal set Θ_x is called Poincaré chaotic if x is an unpredictable point.*

A main result of this chapter is mentioned in the next theorem, and it is valid for both flows and semi-flows on X.

Theorem 2.1.2. [3] *The dynamics on Θ_x is sensitive if $p \in X$ is an unpredictable point.*

If the positive semi-trajectory of an unpredictable point $p \in X$ is contained in a compact subset of X, then Θ_x contains an uncountable set of everywhere dense stable P^+ motions. Additionally, since $T(p)$ is dense in Θ_x the transitivity is also valid in the dynamics.

Definition 2.1.3. [3] *The dynamics on the quasi-minimal set Θ_x is called Poincaré chaotic if p is an unpredictable point.*

In the framework of chaos there may be infinitely many periodic motions. For instance, chaos in the sense of Devaney [15] and Li–Yorke [20] admits a basis consisting of periodic motions. However, in our case Poisson stable motions take place in the dynamics instead of the periodic motions.

Example 2.1.1. [3] *Let us take into account the following space of bi-infinite sequences* [32],

$$\Sigma^2 = \{s = (\ldots s_{-2}s_{-1}.\ s_0s_1s_2\ldots):\ s_j = 0 \text{ or } 1 \text{ for each } j\},$$

with the metric

$$d[s, \bar{s}] = \sum_{k=-\infty}^{\infty} \frac{|s_k - \bar{s}_k|}{2^{|k|}},$$

where $s = (\ldots s_{-2}s_{-1}.\ s_0s_1s_2\ldots)$, $\bar{s} = (\ldots \bar{s}_{-2}\bar{s}_{-1}.\ \bar{s}_0\bar{s}_1\bar{s}_2\ldots) \in \Sigma^2$. The shift map σ: $\Sigma^2 \to \Sigma^2$ is defined as $\sigma(\ldots s_{-2}s_{-1}.\ s_0s_1s_2\ldots) = (\ldots s_{-2}s_{-1}s_0.\ s_1s_2\ldots)$. The map σ is continuous and the metric space Σ^2 is compact [32].

In order to show that the map σ possesses an unpredictable point in Σ^2, we need a collection of finite sequences s_i^m, $m \in \mathbb{N}$, $i = 1, 2, \ldots, 2^m$, consisting of zeros and ones. Let us denote $s_1^1 = (0)$ and $s_2^1 = (1)$. For each $m \in \mathbb{N}$, we recursively define $s_{2i-1}^{m+1} = (s_i^m 0)$, $s_{2i}^{m+1} = (s_i^m 1)$, $i = 1, 2, \ldots, 2^m$. Here, for each m and i, the finite sequences s_{2i-1}^{m+1} and s_{2i}^{m+1} are obtained by, respectively, inserting 0 and 1 at the end of *the sequence* s_i^m *of length m. For example, the sequences with length 2 can be written as* $s_1^2 = (s_1^1 0) = (00)$, $s_2^2 = (s_1^1 1) = (01)$, $s_3^2 = (s_2^1 0) = (10)$, $s_4^2 = (s_2^1 1) = (11)$.

Now, consider the following sequence,

$$s^* = (\ldots s_8^3 s_6^3 s_4^3 s_2^3 s_4^2 s_2^2 .\, s_1^1 s_1^2 s_3^2 s_1^3 s_3^3 s_5^3 s_7^3 \ldots).$$

It was demonstrated in [32] that the trajectory of s^* *is dense in* Σ^2. *We will show that* s^* *is an unpredictable point of the dynamics* (Σ^2, σ). *For each* $n \in \mathbb{N}$, *one can find* $j \in \mathbb{N}$ *such that*

$$s_{2j-1}^{2n+2} = (s_{-n}^* \ldots s_0^* \ldots s_n^* 0)$$

and

$$s_{2j}^{2n+2} = (s_{-n}^* \ldots s_0^* \ldots s_n^* 1).$$

Therefore, there exists a sequence $\{t_n\}$ *with* $t_n \geqslant n + \sum_{k=1}^{2n+1} k 2^{k-1}$, $n \in \mathbb{N}$, *such that* $\sigma^{t_n}(s^*) = s_i^*$ *for* $|\, i\, | \leqslant n$. *Accordingly, the inequality* $d[\sigma^{t_n}(s^*), s^*] \leqslant 1/2^{n-1}$ *is valid so that* $\sigma^{t_n}(s^*) \to s^*$ *as* $n \to \infty$. *Hence,* s^* *is stable* P^+. *In a similar way, one can confirm that* s^* *is stable* P^-. *Note that* Σ^2 *is a quasi-minimal set since* s^* *is Poisson stable. On the other hand, suppose that there exists a natural number n such that*

$$\sigma^{t_n+n+1}(s^*)_i = \sigma^{n+1}(s^*)_i$$

for each $i \geqslant 0$. *Under this assumption we have that* $\sigma^{t_n}(s^*)_i = s_i^*$ *for* $i \geqslant -n$. *This is a contradiction since the sequence* s^* *is not eventually periodic. For this reason, for each* $n \in \mathbb{N}$ *there exists an integer* $\tau_n \geqslant n + 1$ *such that* $\sigma^{t_n+\tau_n}(s^*)_0 \neq \sigma^{\tau_n}(s^*)_0$. *Hence* $d[\sigma^{t_n+\tau_n}(s^*), \sigma^{\tau_n}(s^*)] \geqslant 1$ *for each* $n \in \mathbb{N}$, *and* s^* *is an unpredictable point in* Σ^2.

One of the concepts that has great importance in the theory of dynamical systems is the topological conjugacy, which allows us to make an interpretation of complicated dynamics by using simpler ones. Let X and Y be metric spaces. A flow (semi-flow) f on X is topologically conjugate to a flow (semi-flow) g on Y if there exists a homeomorphism $h: X \to Y$ such that $h \circ f = g \circ h$ [15, 32]. The following theorem can be verified using the arguments presented in [13].

Theorem 2.1.3. *Suppose that X and Y are metric spaces and a flow (semi-flow) f on X is topologically conjugate to a flow (semi-flow) g on Y. If there exists an unpredictable point whose trajectory is contained in a compact subset of X, then there also exists an unpredictable point whose trajectory is contained in a compact subset of Y.*

Example 2.1.2. [3] *(Smale horseshoe) Since the shift map* σ *on* Σ^2 *is topologically conjugate to the Smale horseshoe* [15, 32], *one can conclude by using theorem* 2.1.3 *that the horseshoe map possesses the Poincaré chaos.*

Example 2.1.3. [3] *On the other hand, let us consider the Hénon map*

$$\begin{aligned} x_{n+1} &= \alpha - \beta y_n - x_n^2 \\ y_{n+1} &= x_n, \end{aligned} \tag{2.1}$$

where $\beta \neq 0$ and $\alpha \geqslant (5 + 2\sqrt{5})(1 + |\beta|)^2/4$. It was proved by Devaney and Nitecki [16] *that the map* (2.1) *possesses a Cantor set in which the map is topologically conjugate to the shift map σ on Σ^2. Therefore, theorem* 2.1.3 *also implies the presence of an unpredictable point and a trajectory in the dynamics of equation* (2.1).

Example 2.1.4. [3] *Next, as an example of a semi-flow, consider the following space of infinite sequences* [15],

$$\Sigma_2 = \{s = (s_0 s_1 s_2 \ldots): s_j = 0 \text{ or } 1 \text{ for each } j\}$$

with the metric

$$d[s, \bar{s}] = \sum_{k=0}^{\infty} \frac{|s_k - \bar{s}_k|}{2^k},$$

where $s = (s_0 s_1 s_2 \ldots)$, $\bar{s} = (\bar{s}_0 \bar{s}_1 \bar{s}_2 \ldots) \in \Sigma_2$. The shift map $\sigma: \Sigma_2 \to \Sigma_2$ is defined as $\sigma(s_0 s_1 s_2 \ldots) = (s_1 s_2 s_3 \ldots)$. As in the case of the space of bi-infinite sequences, the metric space Σ_2 is compact and the map σ is continuous [15, 32].

Let us take into account the sequence

$$s^* = (\underbrace{0\ 1}_{1\text{ blocks}}\ |\underbrace{00\ 01\ 10\ 11}_{2\text{ blocks}}|\underbrace{000\ 001\ 010\ 011\ \ldots}_{3\text{ blocks}}|\ldots),$$

which is constructed by successively listing all blocks of zeros and ones of length n, then length $n+1$, etc. This sequence is non-periodic and its trajectory $\mathcal{T}(s^*) = \{\sigma^n(s^*): n = 0, 1, 2, \ldots\}$ is dense in Σ_2 [15] . *Note that the number of all blocks of length n in s^* is 2^n. Based upon the construction of s^*, there exists a sequence $\{t_n\}$ satisfying $t_n \geqslant \sum_{j=1}^{n} j2^j$, $n \in \mathbb{N}$, such that $s_i^* = \sigma^{t_n}(s^*)_i$ for each $i = 0, 1, 2, \ldots, n$. Clearly, $t_n \to \infty$ as $n \to \infty$ and $d[\sigma^{t_n}(s^*), s^*] \leqslant 1/2^n$ so that $\sigma^{t_n}(s^*) \to s^*$ as $n \to \infty$. Hence, s^* is stable P^+. In a very similar way to the bi-infinite sequences in example* 2.1.1, *one can show the existence of a sequence $\{\tau_n\}$, $\tau_n \to \infty$ as $n \to \infty$, such that $d[\sigma^{t_n+\tau_n}(s^*), \sigma^{\tau_n}(s^*)] \geqslant 1$ for each $n \in \mathbb{N}$. Thus s^* is an unpredictable point in Σ_2.*

Example 2.1.5. [3] *It was shown in* [26] *that the logistic map $x_{n+1} = \mu x_n(1 - x_n)$ possesses an invariant Cantor set $\Lambda \subset [0, 1]$, and the map on Λ* is *is topologically conjugate to σ on Σ_2 for $\mu > 4$. Therefore, the map with $\mu > 4$ possesses an unpredictable point and a trajectory in accordance with theorem* 2.1.3.

2.2 Unpredictable functions

This section considers the unpredictability as a phenomenon in functional spaces. The last section considers the property of the dynamics in abstract metric space X, and it is natural to specify the space for applications. This implies that the map of the dynamics has to be specified and thus the shift along the time axis as the map is considered.

Let us denote by $C(\mathbb{R})$ the set of continuous functions defined on $\mathbb{R}$ with values in $\mathbb{R}^m$, and assume that $C(\mathbb{R})$ has the topology of uniform convergence on compact sets, i.e. a sequence $\{h_k\}$ in $C(\mathbb{R})$ is said to converge to a limit h if for every compact set $\mathcal{U} \subset \mathbb{R}$ the sequence of restrictions $\{h_k|_{\mathcal{U}}\}$ converges to $\{h|_{\mathcal{U}}\}$ uniformly. The topology of the convergence on compact sub-intervals can be metricized [27].

One can define a metric ρ on $C(\mathbb{R})$ as

$$\rho(h_1, h_2) = \sum_{k=1}^{\infty} 2^{-k}\rho_k(h_1, h_2), \tag{2.2}$$

where h_1, h_2 belong to $C(\mathbb{R})$ and

$$\rho_k(h_1, h_2) = \min\left\{1, \sup_{s\in[-k,k]}\left\|h_1(s) - h_2(s)\right\|\right\}, \; k \in \mathbb{N}.$$

Let us define the mapping $\pi: \mathbb{R}_+ \times C(\mathbb{R}) \to C(\mathbb{R})$ by $\pi(t, h) = h_t$, where $h_t(s) = h(t + s)$. The mapping π is a semi-flow on $C(\mathbb{R})$ and it is called the Bebutov dynamical system [27].

Using the Bebutov dynamical system we provide the descriptions of a Poisson function and an unpredictable function in the following definitions.

Definition 2.2.1. [27] *A Poisson stable point of the Bebutov dynamical system is said to be a Poisson function.*

Definition 2.2.2. [2] *An unpredictable function is an unpredictable point of the Bebutov dynamical system.*

It is clear that an unpredictable function is a Poisson function. According to theorem III.3 in [27], a motion $\pi(t, h)$ lies in a compact set if h is a bounded and uniformly continuous function. Therefore, an unpredictable function h determines Poincaré chaos in the Bebutov dynamical system if it is bounded and uniformly continuous. Moreover, any system of differential equations which admits uniformly continuous and bounded unpredictable solution has a Poincaré chaos. For differential equations, we say that a solution is an unpredictable one if it is uniformly continuous and bounded on the real axis.

On the basis of the above discussions we shall introduce the following definition, which will be of use in the subsequent chapters of this book.

Definition 2.2.3. [5] *A uniformly continuous and bounded function $v: \mathbb{R} \to \mathbb{R}^p$ is unpredictable if there exist positive numbers ϵ_0, δ and sequences t_n, u_n both of which diverge to infinity such that $v(t + t_n) \to v(t)$ as $n \to \infty$ uniformly on compact subsets of $\mathbb{R}$ and $\|v(t + t_n) - v(t)\| \geqslant \epsilon_0$ for each $t \in [u_n - \delta, u_n + \delta]$ and $n \in \mathbb{N}$.*

To make the research of chaos more friendly to discrete equations, we introduce the concept of an unpredictable sequence as a specific unpredictable function on the set of integers. It is convenient to be verified as a solution of a discrete equation.

The following definition of an unpredictable sequence was first mentioned in [4] as an analog for definition 2.2.3.

Definition 2.2.4. *A bounded sequence* $\{\varphi_n\}$, $n \in \mathbb{Z}$, *in* $\mathbb{R}^p$ *is called unpredictable if there exist a positive number* ϵ_0 *and sequences* $\{\zeta_k\}$, $\{\eta_k\}$, $k \in \mathbb{N}$, *of positive integers, both of which diverge to infinity, such that* $\left\| \varphi_{n+\zeta_k} - \varphi_n \right\| \to 0$ *as* $k \to \infty$ *for each n in bounded intervals of integers and* $\left\| \varphi_{\zeta_k+\eta_k} - \varphi_{\eta_k} \right\| \geqslant \epsilon_0$ *for each* $k \in \mathbb{N}$.

Definition 2.2.4 is of central use in the current book. This is required by the method of the proof. However, in future analyses there may be a need for another definition, which can be considered as a direct specification of definition 2.2.4 as well as an analog of definition 2.2.3.

Definition 2.2.5. *A bounded sequence* $\{\varphi_n\}$, $n \in \mathbb{Z}$, *in* $\mathbb{R}^p$ *is called unpredictable if there exist a positive number* ϵ_0 *and sequences* $\{\zeta_k\}$, $\{\eta_k\}$, $k \in \mathbb{N}$, *of positive integers, both of which diverge to infinity, such that* $\left\| \varphi_{\zeta_k} - \varphi_0 \right\| \to 0$, *as* $k \to \infty$, *and* $\left\| \varphi_{\zeta_k+\eta_k} - \varphi_{\eta_k} \right\| \geqslant \epsilon_0$ *for each* $k \in \mathbb{N}$.

The topologies in definitions 2.2.3 and 2.2.4 are metrizable [27]. Consequently, the existence of an unpredictable sequence in the sense of definition 2.2.4 indicates the presence of Poincaré chaos [2, 3]. In what follows, an unpredictable sequence and an unpredictable solution are understood as mentioned in definition 2.2.4.

Next we will show the presence of an unpredictable point in the symbolic dynamics (Σ_2, σ, d) of example 2.1.4 with a distinguishing feature. In the example the sequence s^* is proved to be unpredictable. In the proof of the next lemma an element $s^{**} = (s_0^{**} s_1^{**} s_2^{**} \ldots)$ of Σ_2 will be constructed in a similar way to s^* with the only difference that the order of the blocks will be chosen in a special way. An extension of the constructed sequence to the left-hand side will also be provided.

Lemma 2.2.1. [2] *For each increasing sequence* $\{m_n\}$ *of positive integers, there exist a sequence* $s^{**} \in \Sigma_2$ *and sequences* $\{\alpha_n\}$, $\{\beta_n\}$ *of positive integers, both of which diverge to infinity, such that*

(i) $d(\sigma^{\alpha_n+r}(s^{**}), \sigma^r(s^{**})) \leqslant 2^{-m_n}$, $r = -n, -n+1, \ldots, n$,

(ii) $d(\sigma^{\alpha_n+\beta_n}(s^{**}), \sigma^{\beta_n}(s^{**})) \geqslant 1$ *for each* $n \in \mathbb{N}$.

Proof. Fix an arbitrary increasing sequence $\{m_n\}$ of positive integers. For each $n \in \mathbb{N}$, define $\alpha_n = \sum_{k=1}^{n+m_n} k 2^k$ and $\beta_n = n + m_n + 1$. Clearly both of the sequences $\{\alpha_n\}$ and $\{\beta_n\}$ diverge to infinity. We will construct a sequence $s^{**} \in \Sigma_2$ such that the inequalities (i) and (ii) are valid.

First we choose the terms $s_0^{**}, s_1^{**}, \ldots, s_{\alpha_1-1}^{**}$ by successively placing the blocks of zeros and ones in increasing length, starting from the blocks of length 1 and ending with the ones with length $m_1 + 1$. The order of the blocks with the same length can be arbitrary without any repetitions. Let us take $s^{**}_{\alpha_1+k} = s^{**}_k$ for $k = 0, 1, \dots, m_1 + 1$, i.e. the terms of the first block $(s^{**}_{\alpha_1} s^{**}_{\alpha_1+1} \cdots s^{**}_{\alpha_1+m_1+1})$ of length $m_1 + 2$ is chosen to be the same as the first $m_1 + 2$ terms of the sequence s^{**}. Moreover, we take the second block of length $m_1 + 2$ in a such a way that its first term $s^{**}_{\alpha_1+\beta_1}$ is different from $s^{**}_{\beta_1}$. After that we continue placing the remaining blocks of length $m_1 + 2$ and the ones with length greater than $m_1 + 2$ until the last block of length $m_2 + 2$ and, again, the blocks of the same length can be in any order without repetition.

For each $n \geqslant 2$ we set the last block of length $m_n + n$ such that $s^{**}_{\alpha_n-k} = s^{**}_{\alpha_k-k}$ for each $k = 1, 2, \dots, n-1$. Then the terms of the first block of length $m_n + n + 1$ are constituted by taking $s^{**}_{\alpha_n+k} = s^{**}_k$, $k = 0, 1, \dots, m_n + n$. Moreover, the second block of length $m_n + n + 1$ is chosen such that $s^{**}_{\alpha_n+\beta_n} \neq s^{**}_{\beta_n}$. Finally, the remaining blocks of zeros and ones are successively placed, similar to the case mentioned above, so that the lengths of the blocks in s^{**} are in increasing order and there are no repetitions within the blocks of the same length. In this way the construction of the sequence $s^{**} \in \Sigma_2$ is completed. We fix the extension of the sequence s^{**} to the left by choosing $\sigma^{-k}(s^{**})_0 = s^{**}_{\alpha_k-k}$, $k \in \mathbb{N}$. For each $n \in \mathbb{N}$ we have $\sigma^{\alpha_n-n}(s^{**})_k = \sigma^{-n}(s^{**})_k$, $k = 0, 1, \dots, m_n + 2n$ and $\sigma^{\alpha_n+\beta_n}(s^{**})_0 \neq \sigma^{\beta_n}(s^{**})_0$ so that the inequalities (i) and (ii) are valid.

Through the technique in [15] one can show that the trajectory $T(s^{**}) = \{\sigma^i(s^{**})\colon i \in \mathbb{Z}\}$ is dense in Σ_2, i.e. $\Theta_{s^{**}} = \Sigma_2$. By lemma 2.2 in [3] any sequence $\sigma^i(s^{**})$, $i \in \mathbb{Z}$, is an unpredictable point of the Bernoulli dynamics on Σ_2 and Σ_2 is a quasi-minimal set. It implies from the last theorem that $T(s^{**})$ is an unpredictable function on $\mathbb{Z}$, i.e. an unpredictable sequence. According to theorem 3.1 presented in [3], the dynamics on Σ_2 is Poincaré chaotic. Moreover, there are infinitely many unpredictable sequences in the set.

Example 2.2.1. [2] *We will demonstrate the presence of an unpredictable solution of the equation*

$$\eta_{n+1} = F_\mu(\eta_n), \tag{2.3}$$

where $F_\mu(s) = \mu s(1 - s)$ is the logistic map. More precisely, it is proved that for each $\mu \in [3 + (2/3)^{1/2}, 4]$ and sequence of positive numbers $\delta_n \to 0$, there exists a solution $\{\eta_n\}$, $n \in \mathbb{Z}$, of equation (2.3) *such that*

(i) $\left| \eta_{i_n+r} - \eta_r \right| < \delta_n$, $r = -h_0 n, -h_0 n + 1, \dots, h_0 n$,

(ii) $\left| \eta_{i_n+j_n} - \eta_{j_n} \right| \geqslant \epsilon_0$ *for each* $n \in \mathbb{N}$,

where ϵ_0 is a positive number, $h_0 > 4$ is a natural number and $\{i_n\}$, $\{j_n\}$ are integer valued sequences both of which diverge to infinity.

Fix $\mu \in [3 + (2/3)^{1/2}, 4]$ and a sequence $\{\delta_n\}$ of positive real numbers with $\delta_n \to 0$ as $n \to \infty$. Take a neighborhood $U \subset [0, 1]$ of the point $1 - 1/\mu$. According to theorem 6 of [28], *there exist a natural number $h_0 > 4$ and a Cantor set $\Lambda \subset U$ such that the* *map* $F_\mu^{h_0}$ *on* Λ *is topologically conjugate to the Bernoulli shift* σ *on* Σ_2. *Therefore, there exists a homeomorphism* $S\colon \Sigma_2 \to \Lambda$ *such that* $S\circ\sigma = F_\mu^{h_0}\circ S$. *Since* S *is uniformly continuous on* Σ_2, *for each* $n \in \mathbb{N}$, *there exists a number* $\bar{\delta}_n > 0$ *such that for each* s^1, $s^2 \in \Sigma_2$ *with* $d(s^1, s^2) < \bar{\delta}_n$, *we have* $|\eta^1 - \eta^2| < \delta_n/\mu^{h_0-1}$, *where* $\eta^1 = S(s^1)$, $\eta^2 = S(s^2)$.

Let $\{m_n\}$ *be an increasing sequence of natural numbers such that* $2^{-m_n} < \bar{\delta}_n$ *for each* $n \in \mathbb{N}$. *According to lemma 2.2.1, there exist a sequence* $s^{**}\in\Sigma_2$ *and sequences* $\{\alpha_n\}$, $\{\beta_n\}$, *both of which diverge to infinity such that* $d(\sigma^{\alpha_n+r}(s^{**}), \sigma^r(s^{**})) \leqslant 2^{-m_n}$, $r = -n, -n+1, \dots, n$, *and* $d(\sigma^{\alpha_n+\beta_n}(s^{**}), \sigma^{\beta_n}(s^{**})) \geqslant 1$ *for each* $n \in \mathbb{N}$.

Now, let $\{\eta_n\}$, $n \in \mathbb{Z}$, *be the solution of equation* (2.3) *with* $\eta_{h_0k} = S(\sigma^k(s^{**}))$, $k \in \mathbb{Z}$. *Since the inequality* $\left| F_\mu(u_1) - F_\mu(u_2)\right| \leqslant \mu \mid u_1 - u_2 \mid$ *is valid for every* $u_1, u_2 \in [0, 1]$, *we have for each* $n \in \mathbb{N}$ *that* $\left| \eta_{i_n+r} - \eta_r \right| < \delta_n$, $r = -h_0n, -h_0n+1, \dots, h_0n$, *where* $i_n = h_0\alpha_n$. *In addition, using the arguments presented in* [13], *one can verify the existence of a positive number* ϵ_0 *such that* $\left| \eta_{i_n+j_n} - \eta_{j_n} \right| \geqslant \epsilon_0$, $n \in \mathbb{N}$, *where* $j_n = h_0\beta_n$.

Example 2.2.2. *According to the last example there exists the unpredictable solution* ψ_i, $i \in \mathbb{Z}$, *of the logistic discrete equation*

$$\lambda_{i+1} = \mu\lambda_i(1 - \lambda_i), \tag{2.4}$$

with $\mu = 3.92$. *The sequence belongs to the unit interval* [0, 1]. *There exist a positive number* ϵ_0 *and sequences* ζ_n, η_n *both of which diverge to infinity such that* $|\psi_{i+\zeta_n} - \psi_i| \to 0$ *as* $n \to \infty$ *for each* i *in bounded intervals of integers and* $|\psi_{\zeta_n+\eta_n} - \psi_{\eta_n}| \geqslant \epsilon_0$ *for each* $n \in \mathbb{N}$.

Consider the following integral

$$\Theta(t) = \int_{-\infty}^{t} e^{-2(t-s)}\Omega(s)ds,\ t \in \mathbb{R}, \tag{2.5}$$

where $\Omega(t)$ *is a piecewise constant function defined on the real axis through the equation* $\Omega(t) = \psi_i$ *for* $t \in [i, i+1)$, $i \in \mathbb{Z}$.

It is worth noting that $\Theta(t)$ *is bounded on the whole real axis such that* $\sup_{t\in\mathbb{R}} \mid \Theta(t)| \leqslant 1/2$.

Next we will show that $\Theta(t)$ *is an unpredictable scalar function.*

Consider a fixed bounded and closed interval $[\alpha, \beta]$ *of the axis and a positive number* ϵ. *Without loss of generality one can assume that* α *and* β *are integers. Let us fix a positive number* ξ *and an integer* $\gamma < \alpha$, *which satisfy the following inequalities* $e^{-2(\alpha-\gamma)} < \frac{\epsilon}{2}$ *and* $\xi[1 - e^{-2(\beta-\gamma)}] < \epsilon$. *Let* n *be a large natural number such that* $\mid \Omega(t+\zeta_n) - \Omega(t)| < \xi$ *on* $[\gamma, \beta]$. *Then for all* $t \in [\alpha, \beta]$ *we obtain that*

$$|\Theta(t+\zeta_n) - \Theta(t)| = |\int_{-\infty}^{t} e^{-2(t-s)}(\Omega(s+\zeta_n) - \Omega(s))ds| =$$

$$|\int_{-\infty}^{\gamma} e^{-2(t-s)}(\Omega(s+\zeta_n) - \Omega(s))ds + \int_{\gamma}^{t} e^{-2(t-s)}(\Omega(s+\zeta_n) - \Omega(s))ds| \leqslant$$

$$\int_{-\infty}^{\gamma} e^{-2(t-s)}2ds + \int_{\gamma}^{\beta} e^{-2(t-s)}\xi ds \leqslant e^{-2(\alpha-\gamma)} + \frac{\xi}{2}[1 - e^{-2(\beta-\gamma)}] < \frac{\epsilon}{2} + \frac{\epsilon}{2} = \epsilon.$$

Thus $|\Theta(t+\zeta_n)-\Theta(t)| \to 0$ *as* $n\to\infty$ *uniformly on the interval* $[\alpha,\beta]$. *Moreover, the following inequalities are valid,* $|\Omega(t+\zeta_n)-\Omega(t)| = |\psi_{\zeta_n+\eta_n}-\psi_{\eta_n}| \geqslant \epsilon_0$, $t\in[\eta_p,\eta_p+1)$, $p\in\mathbb{N}$.

Let us fix the numbers κ *and* n *and consider two alternative cases: (i)* $|\Theta(\eta_n+\zeta_n)-\Theta(\eta_n)| < \frac{\epsilon_0}{8}$ *and (ii)* $|\Theta(\eta_n+\zeta_n)-\Theta(\eta_n)| \geqslant \frac{\epsilon_0}{8}$.

(i) There exists a positive number $\kappa<1$ *such that* $e^{-2\kappa}=\frac{2}{3}$. *Using the relation*

$$\Theta(t+\zeta_n)-\Theta(t)=\Theta(\eta_n+\zeta_n)-\Theta(\eta_n)+\int_{\eta_n}^{t} e^{-2(t-s)}(\Omega(s+\zeta_n)-\Omega(s))ds, \quad (2.6)$$

we obtain that

$$|\Theta(t+\zeta_n)-\Theta(t)| \geqslant |\int_{\eta_n}^{t} e^{-2(t-s)}|\Omega(s+\zeta_n)-\Omega(s))|ds - |\Theta(\eta_n+\zeta_n)-\Theta(\eta_n)| \geqslant$$

$$\int_{\eta_n}^{t} e^{-2(t-s)}\epsilon_0 ds - \frac{\epsilon_0}{8} \geqslant \frac{\epsilon_0}{2}(1-e^{-2\kappa}) - \frac{\epsilon_0}{8} = \frac{\epsilon_0}{24},$$

for $t\in[\eta_n+\kappa,\eta_n+1)$.

(ii) There exists a positive number $\kappa<1$ *such that* $1-e^{-2\kappa}=\frac{\epsilon_0}{12}$. *From the relation* (2.6) *we obtain*

$$|\Theta(t+\zeta_n)-\Theta(t)| \geqslant |\Theta(\eta_n+\zeta_n)-\Theta(\eta_n)| - |\int_{\eta_n}^{t} e^{-2(t-s)}(\Omega(s+\zeta_n)-\Omega(s))ds| \geqslant$$

$$\frac{\epsilon_0}{8} - \int_{\eta_n}^{t} e^{-2(t-s)}2ds \geqslant \frac{\epsilon_0}{8} - [1-e^{-2\kappa}] = \frac{\epsilon_0}{24},$$

for $t\in[\eta_n,\eta_n+\kappa)$.

Thus $\Theta(t)$ *is an unpredictable function.*

2.3 Notes

Emphasizing the ingredients of Devaney [15] and Li–Yorke [20] chaos, one can see that the definitions of chaos have been considered by means of sets of motions, but not through a single description of motion. In our papers [2, 3, 6], for the first time in the literature, chaos is initiated from a single motion, which has sophisticated behavior. Thus the line of equilibrium, periodic function, quasi-periodic function, almost periodic function, recurrent function and Poisson stable function is extended with the new element—unpredictable motion. This supplement to the line creates the possibility of other functions behind the known ones.

An essential point to discuss is the sensitivity or unpredictability. In the literature sensitivity has been considered through initially nearby different motions. However, we talk about unpredictability as an interior property of a single trajectory. Then chaos appears in a neighborhood of the trajectory. The symbolic dynamics illustrates all the results and it is an important tool for the investigation of the complicated dynamics of continuous-time systems such as the Lorenz and Rössler equations [23]. It is worth noting that unpredictable points can be replicated by the techniques summarized in [1].

One can see the proximity of chaos and quasi-minimal sets by comparing their definitions [15, 20, 24]. Transitivity is a common feature of them and the closure of a Poisson stable trajectory contains infinitely many Poisson stable orbits. In its own turn we know that a periodic trajectory is also Poisson stable. Possibly, the continuum of Poisson stable trajectories is an ultimate form of infinitely many cycles known for a chaos. One may also ask whether sensitivity is proper for any quasi-minimal set. These are the questions to discover more relations between quasi-minimal sets and chaos.

We utilize the topology of uniform convergence on any compact subset of the real axis to introduce the unpredictable functions. More precisely, the Bebutov dynamical system [27] has been applied. All these make our task of incorporating the investigation of chaos into the theory of differential equations, initiated in papers [4, 5, 7] and the book [6], appear to proceed in the correct way.

In the instrumental sense, discreteness has been the main object in chaos investigation. To check this it is sufficient to recall the definitions of chaos [15, 20, 32], which are based on sequences and maps, as well as the Smale horseshoe and logistic maps, and Bernoulli shift [32], which are at the core of the chaos theory. One can say that the stroboscopic observation of motion was the single way to indicate irregularity in a continuous dynamics. Embedding the research in the theory of differential equations requires definitions of chaos for continuous dynamics, which are not related to discreteness [1]. The research as well as the origin of the chaos [25] gave us strong arguments for the development of motions in classical dynamical systems theory [14] by proceeding behind Poisson stable points to unpredictable points [3]. Then the dynamics has been specified such that a function that is bounded on the real axis is an unpredictable point [5, 7]. In our papers we have demonstrated that unpredictable functions are easy to analyze as solutions of differential equations. This paradigm is not completed if one does not consider discrete equations. This is the reason why we deliver discrete analogs for unpredictable functions, calling them unpredictable sequences. The role of the discrete unpredictability cannot be underestimated for applications as well as for theoretical analysis, in particular for the modern development of computer technologies, software and robotics [17, 30].

To protect the reader from misunderstanding, we will quote the following from the reviewer MR3505917 of our results in [3]: 'We note, however, that these results for a particular trajectory and semi-trajectory are covered by the following rather general result that is easy to prove: Assume X is a compact metric space and consider either a cascade generated by an expansive homeomorphism $X \to X$ or a semi-cascade generated by an expanding map $X \to X$. Then if a point in X is stable P^+ and is not periodic, it is unpredictable.'

Let us recall that a map $f: X \to X$ is expansive if there exists $\epsilon_0 > 0$ such that, for any $x, y \in X, x \neq y$, there exists n such that $d(f^n x, f^n y) > \epsilon_0$ [15] and the sensitivity does not imply the expansiveness. This is why the suggested assertion cannot cover theorem 2.1.1 and the statement of the reviewer is not correct. Otherwise, the reviewer is invited to clarify his/her position. At the same time there are already serious studies which confirm that the subject of our research has very interesting prospects [22, 31].

References

[1] Akhmet M and Fen M O 2016 *Replication of Chaos in Neural Networks, Economics and Physics* (Berlin: Springer)

[2] Akhmet M and Fen M O 2016 Poincaré chaos and unpredictable functions *Commun. Nonlinear Sci. Numer. Simul.* **48** 85–94

[3] Akhmet M and Fen M O 2016 Unpredictable points and chaos *Commun. Nonlinear Sci. Numer. Simul.* **40** 1–5

[4] Akhmet M and Fen M O 2017 Existence of unpredictable solutions and chaos *Turk. J. Math.* **41** 254–66

[5] Akhmet M and Fen M O 2018 Non-autonomous equations with unpredictable solutions *Commun. Nonlinear Sci. Numer. Simul.* **59** 657–70

[6] Akhmet M, Fen M O and Alejaily E M 2020 *Dynamics with Chaos and Fractals* (Berlin: Springer)

[7] Akhmet M, Fen M O, Tleubergenova M and Zhamanshin A 2019 Unpredictable solutions of linear differential and discrete equations *Turk. J. Math.* **43** 2377–89

[8] Akhmet M and Fen M O 2014 Entrainment by chaos *J. Nonlinear Sci.* **24** 411–39

[9] Akhmet M and Fen M O 2015 Attraction of Li–Yorke chaos by retarded SICNNs *Neurocomputing* **147** 330–42

[10] Akhmet M U 2009 Creating a chaos in a system with relay *Int. J. Qual. Theory Differ. Equ. Appl.* **3** 3–7

[11] Akhmet M U 2009 Dynamical synthesis of quasi-minimal sets *Int. J. Bifurcat. Chaos* **19** 2423–7

[12] Akhmet M U 2009 Shadowing and dynamical synthesis *Int. J. Bifurcat. Chaos* **19** 3339–46

[13] Banks J, Brooks J, Cairns G, Davis G and Stacey P 1992 On Devaney's definition of chaos *Am. Math. Mon.* **99** 332–4

[14] Birkhoff G D 1991 *Dynamical Systems* (Providence, RI: Colloquium)

[15] Devaney R L 1989 *An Introduction to Chaotic Dynamical Systems* (Canada; USA: Addison-Wesley)

[16] Devaney R L and Nitecki Z 1979 Shift automorphisms in the Henon mapping *Math. Phys.* **67** 137–46

[17] Ditto W L, Murali K and Sinha S 2008 Chaos computing: ideas and implementations *Philos. Trans. R. Soc.* A **366** 653–64

[18] Hartman P 2002 *Ordinary Differential Equations* (New York: SIAM)

[19] Hilmy H 1936 Sur les ensembles quasi-minimaux dans les systèmes dynamiques *Ann. Math.* **37** 899–907

[20] Li T Y and Yorke J A 1975 Period three implies chaos *Am. Math. Mon.* **82** 985–92

[21] Lorenz E N 1984 The local structure of chaotic attractor in four dimensions *Physica* D **13** 90–104

[22] Miller A 2019 Unpredictable points and stronger versions of Ruelle–Takens and Auslander–Yorke chaos *Topol. Appl.* **253** 7–16

[23] Mischaikow K and Mrozek M 1995 Chaos in the Lorenz equations: a computer assisted proof *Bull. AMS* **32** 66–72
[24] Nemytskii V V and Stepanov V V 1960 *Qualitative Theory of Differential Equations* (Princeton, NJ: Princeton University Press)
[25] Poincaré H 1957 *New Methods of Celestial Mechanics* vol 1–3 (New York: Dover)
[26] Robinson C 1995 *Dynamical Systems: Stability, Symbolic Dynamics, and Chaos* (Boca Raton, FL: CRC Press)
[27] Sell G R 1971 *Topological Dynamics and Ordinary Differential Equations* (London: Van Nostrand Reinhold)
[28] Shi Y and Yu P 2007 On chaos of the logistic maps *Dyn. Contin. Discrete Impuls. Syst.* B **14** 175–95
[29] Shilnikov L 1997 Homoclinic chaos *Nonlinear Dynamics, Chaotic and Complex Systems* (Cambridge: Cambridge University Press) pp 39–63
[30] Staingrube S, Timme M, Worgotter F and Mannonpong P 2010 Self-organized adaptation of simple neural circuits enables complex robot behavior *Nat. Phys.* **6** 224–30
[31] Thakur R and Das R 2020 Strongly Ruelle–Takens, strongly Auslander–Yorke and Poincaré chaos on semiflows *Commun. Nonlinear Sci. Numer. Simul.* **81** 105018
[32] Wiggins S 1988 Global bifurcation and chaos *Analytical Methods* (New York: Springer)

Chapter 3

Domain structured dynamics: theory

The method of investigation of a dynamics by projection on Bernoulli shift orbits, along infinite sequences of symbols, is a most powerful tool for the theoretical study of chaos. However, we feel that there is a strong deficiency—one has to establish the projection, which is an auxiliary mapping, and is evidently not an easy task. This is because 'to establish' means not only the search for the projection as a concrete map, but also verification that it is diffeomorphism/homeomorphism, and this is an artform similar to the construction of Lyapunov functions. What we are suggesting is to structure a state space properly such that the rule of switching between the labeled elements reflects the desired dynamical features directly. Sufficient background can be found in the current research. It is built of clearly formulated and easy-to-verify diameter and separation conditions. It is hoped that the method of domain structured dynamics will be developed further and other fruitful conditions will be suggested to observe much more interesting phenomena. One can easily see that we verify the searched dynamics transparently, without 'mirroring' it in a symbolic dynamics. This is what we promised as an easy way of learning about chaos. This will be seen theoretically in the current chapter and in the next chapter through applications. Surprisingly, it is possible to simplify the proof of chaos in classical models. Moreover, the method appears to be universal to prove chaos in fractals. This is predictable, since self-similarity is in the basis of the definition of classical fractals. Moreover, by using the domain structured chaos we extract the dynamical essence of hyperbolic sets. In our recent research abstract hyperbolic sets are newly proposed for a clear and common vision of chaos in many models in the area. This time we consider the Smale horseshoe and baker transformation, and suggest considering Anosov automorphisms, Plykin attractors and other similar constructions for prospective analysis. In the following chapters we will discuss the benefits arising from our proposals for bringing chaotic and stochastic motions closer. Necessary chaotic features will be discovered in such dynamics as Bernoulli schemes and Markov chains in the following chapters. Finally, for use in science we

doi:10.1088/978-0-7503-3507-2ch3

suggest a new concept of modular chaos, when motions are whirling among various and different structured domains. This generates new opportunities to investigate unbounded motions and random processes with continuous time.

3.1 Basic definitions and theorems

In the first part of this section we will provide the basis of the domain structured dynamics—abstract similarity sets. They are applied for domain structured dynamics through a labeling method. Since the labeling is applied to dynamics we will say that a state space is structured or a domain is structured. The labeling can be performed with multiplicity one, then the domain is an uncountable set. To make structuring flexible in applications one can assign multiple labels to a single element, that is, several and even infinitely many labels. Correspondingly, a state space may consist of countable and even finite number elements. This is convenient, for example, to analyze random processes with finite state spaces.

3.1.1 Abstract similarity sets

We shall consider a collection

$$\mathcal{F} = \left\{\mathcal{F}_{i_1 i_2 \ldots i_n \ldots} \colon i_k = 1, 2, \ldots, m, \; k = 1, 2, \ldots\right\}, \tag{3.1}$$

where m is a natural number and members are called *labels*. The set $\mathcal{F}$ is called the *abstract similarity set* in what follows. Another name, the *abstract label set* or simply the *label set* is also sensible for the set $\mathcal{F}$, considering application for domain structured dynamics.

For dynamical needs we shall consider the following subsets, *n-dimensional rectangles*,

$$\mathcal{F}_{i_1 i_2 \ldots i_n} = \bigcup_{j_k = 1,2,\ldots,m} \mathcal{F}_{i_1 i_2 \ldots i_n j_1 j_2 \ldots}, \tag{3.2}$$

where indices $i_1, i_2, \ldots, i_n$, are fixed.

It is not difficult to see that

$$\mathcal{F} \supseteq \mathcal{F}_{i_1} \supseteq \mathcal{F}_{i_1 i_2} \supseteq \cdots \supseteq \mathcal{F}_{i_1 i_2 \ldots i_n} \supseteq \mathcal{F}_{i_1 i_2 \ldots i_n i_{n+1}} \ldots, \quad i_k = 1, 2, \ldots, m, \;\; k = 1, 2, \ldots.$$

That is, the rectangles are united as a nested sequence. Thus we have completed comprehension of the label set, if we do not need a metric for a dynamical application.

3.1.2 Abstract similarity map

In this section we concern ourselves with the *self-similarity* of the abstract label set introduced in the last subsection. It is obvious that a self-similarity can be defined only by a mapping and metrical properties are also desirable. We specify the abstract label set as a self-similar set and as a collection of points in a metric space, where it can be considered as a union of infinite shrinking sets with a notation that allows dynamics to be introduced in the set and then one can prove chaos, which is the most important result of the current research. For this purpose a specific map over the invariant self-similar set is defined. The map acts as the identity if the whole set is taken as an argument for the map. This feature is equivalent to the property of self-similarity in ordinary fractals and this is the reason for calling this transformation a similarity map.

We expect that the concept of abstract similarity will create new fundamentals for investigations of chaos and fractals, and we hope it will be a helpful tool in other fields such as harmonic analysis, discrete mathematics, probability and operator algebras [38]. Let us recall that self-similarity can be observed in many problems arising in various areas such as wavelets, fractals and graph systems [38].

Our approach with respect to chaos is characterized by the priority of the domain over the map [6]. More precisely, the usual construction of chaos starts with the description of a map with certain properties such as unimodality, hyperbolicity, period three and topological conjugacy to a standard chaotic map. Next, chaotic behavior is discovered and, finally, the structure of the chaotic attractor is analyzed. For example, in Li–Yorke chaos a map is first defined such that it has period three property and then the domain for the chaos is characterized by specific properties. The same can be said for the chaos of unimodal maps, when the domain of chaos is a Cantor set. In the Smale horseshoe case we can conclude that the map and the domain are simultaneously determined so that the construction of the domain began with an initial set and it is developed step by step using a particular map. Therefore the structure of the domain and the nature of the map are mutually dependent on each other. For the chaos in symbolic dynamics the domain is primarily described as an infinite sequences of symbols, then the map is introduced as a shift on the space. However, the map still has priority since the properties of the sequences are described with respect to the map. In the proposed approach we first construct a domain in a metric space with specific conditions to be a suitable venue for manifestations of chaos. Thereafter the similarity map is built on the basis of the invariance and self-similarity properties of the domain to define an abstract motion. This is why the map is appropriate for an abstract self-similar set as well as for any fractal constructed through self-similarity, thus proving chaos for these classes of fractals becomes possible. Moreover, we drop the continuity requirement for the motion since the chaotic map need not be continuous [1, 41]. In [41], for example, the authors ignore the continuity of some chaotic maps during the discussion of chaos conditions on the product of semi-flows. We regard the continuity of a chaotic map as an important property only from the analytical side, that is to say, it is very useful for handling the map to prove the presence of chaos [61], however, it is not rigorously correct to consider it as an intrinsic property for chaos. Despite the discontinuity of the similarity map, opposite to our desire, one can recognize that the presence of sensitivity and the irregular behavior of simulations make the discussion valuable for theoretical investigations as well as for future applications. As implementations we consider different types of chaos, namely Devaney, Li–Yorke and Poincaré for self-similar fractals, Sierpinski fractals, the Koch curve as well as the Cantor set.

The concept of abstract similarity, in different forms, has been realized for several problems in our papers [5–7].

Let us consider the label set, $\mathcal{F}$, of the last subsection, and determine the map φ: $\mathcal{F} \to \mathcal{F}$ such that

$$\varphi(\mathcal{F}_{i_1 i_2 \ldots i_n \ldots}) = \mathcal{F}_{i_2 i_3 \ldots i_n \ldots}. \tag{3.3}$$

Considering iterations of the map, applied to an n-dimensional rectangle, one can verify that

$$\varphi^n(\mathcal{F}_{i_1 i_2 \ldots i_n}) = \mathcal{F}, \tag{3.4}$$

for arbitrary natural number n and $i_k = 1, 2, \ldots, m$, $k = 1, 2, \ldots, n$. The relations (5.22) and (5.23) give us a reason to call φ the *similarity map* and the number n the *order of similarity*. It is reasonable to call φ the *abstract similarity map* as it is in the abstract similarity set $\mathcal{F}$.

3.1.3 Abstract similarity dynamics

Let a compact metric space (F, d) be given with the distance d. We suggest that the set F can be finite, countable or uncountable. We will say that an abstract similarity set $\mathcal{F}$ is a *label set* for F, if each element of F is labeled at least by one member of the set $\mathcal{F}$ and each element of $\mathcal{F}$ presents a member from F. Thus we do not propose that the uniqueness is required for the relation. That is, if $\mathcal{F}_{i_1 i_2 \ldots i_n \ldots}$ and $\mathcal{F}_{j_1 j_2 \ldots j n, \ldots}$ present the same element f in F, it is not necessary that $i_n = j_n$ for all $n = 1, 2, \ldots$. Moreover, we keep the function d for the set $\mathcal{F}$ considering $d(\mathcal{F}_{i_1 i_2 \ldots i_n \ldots}, \mathcal{F}_{j_1 j_2 \ldots j n, \ldots}) = d(f_1, f_2)$, if $\mathcal{F}_{i_1 i_2 \ldots i_n \ldots}$ and $\mathcal{F}_{j_1 j_2 \ldots j n, \ldots}$ are labels of elements f_1 and f_2 of the set F, respectively, such that $d(\mathcal{F}_{i_1 i_2 \ldots i_n \ldots}, \mathcal{F}_{j_1 j_2 \ldots j n, \ldots}) = 0$ for different labels of the same point in F. If there is uniqueness in the labeling the couple $(\mathcal{F}, d)$ is also a metric space. It is clear that in what follows we can accept the terms *label set* and *abstract similarity set* as synonyms. Assume now that the metric space (F, d) is a domain for a map ψ such that (F, ψ, d) is a dynamical system. We shall say that the maps ψ and φ agree, if for each element f of F, if $\mathcal{F}_{i_1 i_2 \ldots i_n \ldots}$ is a label for f means that $\mathcal{F}_{i_2 \ldots i_n \ldots}$ is a label for $\psi(f)$. This will be called the *image condition* in what follows. One cannot speak of an equivalence for the dynamics, since there is no one-to-one correspondence for the labeling. Otherwise we can call it isometry. The image condition is assumed throughout the book, if the opposite is not specified.

The triple $(\mathcal{F}, d, \varphi)$ is said to be the *abstract similarity dynamics* induced by the dynamical system (F, ψ, d) and it is also said to be domain structured dynamics for the dynamical system. It is important that we call the triple $(\mathcal{F}, d, \varphi)$ a dynamics, but not a dynamical system, since the labeling may not be a bijection. Finally, if a metric space (F, d) is given without any dynamics, we call the triple $(\mathcal{F}, d, \varphi)$ a *domain structured dynamics* in the space (F, d) under the condition that $\mathcal{F}$ is a label set for F.

We will say that for the rectangles $\mathcal{F}_{i_1 i_2 \ldots i_n}$ the *diameter condition* is valid if

$$\sup_{i_k = 1, 2, \ldots, m} \operatorname{diam} \mathcal{F}_{i_1 i_2 \ldots i_n} \to 0 \quad \text{as} \quad n \to \infty, \tag{3.5}$$

where $\operatorname{diam}(A) = \sup\{d(\mathbf{x}, \mathbf{y}) \colon \mathbf{x}, \mathbf{y} \in A\}$ for a set A in $\mathcal{F}$.

In the next example of our study and in future studies, it is important to find an abstract self-similar set for a given model.

Example 3.1.1. [7] *Let us consider the space of symbolic strings of* 0 *and* 1 [26, 61], *which is defined by*

$$\Sigma = \{s_1 s_2 s_3 \ldots : s_k = 0 \text{ or } 1\}.$$

The distance in Σ *is defined by*

$$d(s, t) = \sum_{k=1}^{\infty} \frac{|s_k - t_k|}{2^{k-1}}, \tag{3.6}$$

where $s = s_1 s_2 \ldots$ *and* $t = t_1 t_2 \ldots$ *are two elements in* Σ.

Considering the pattern of the self-similar set, we denote the elements of the set by $\mathcal{S}_{s_1 s_2 \ldots} = s_1 s_2 \ldots$, *and describe the nth order rectangles of strings in* $\mathcal{S}$ *by*

$$\mathcal{S}_{s_1 s_2 \ldots s_n} = \{s_1 s_2 \ldots s_n s_{n+1} s_{n+2} \ldots : s_k = 0 \text{ or } 1\},$$

where $s_1, s_2, \ldots, s_n$ *are fixed symbols. One can show that the distance between any two elements of* $\mathcal{S}_{s_1 s_2 \ldots s_n}$ *is less than* $\frac{1}{2^{n-1}}$. *Moreover,* $d(s_1 s_2 \ldots s_n 000\ldots, s_1 s_2 \ldots s_n 111\ldots) = \frac{1}{2^{n-1}}$. *Therefore,* $\mathrm{diam}(\mathcal{S}_{s_1 s_2 \ldots s_n}) = \frac{1}{2^{n-1}}$. *Consequently,*

$$\lim_{n\to\infty} \mathrm{diam}\, \mathcal{S}_{s_1 s_2 \ldots s_n} = 0,$$

and the diameter condition holds.

The similarity map for the space is the Bernoulli shift, $\sigma(s_1 s_2 s_3 \ldots) = s_2 s_3 s_4 \ldots$. *That is,*

$$\varphi(\mathcal{S}_{s_1 s_2 s_3 \ldots}) = \sigma(s_1 s_2 s_3 \ldots).$$

Thus $(\mathcal{S}, \varphi, d)$ *is an abstract similarity dynamics for the symbolic dynamical system* (Σ, σ, d).

This is a purely illustrative example since it makes us perceive how self-similarity can be defined for abstract objects which are not necessarily geometrical ones. The space of symbolic strings on two symbols has been considered, since it is the most basic example that is frequently used to describe the dynamics on symbolic spaces. However, more generally, the space on m symbols can also be considered.

3.1.4 Domain structured chaos

Let the abstract similarity dynamics $(\mathcal{F}, d, \varphi)$ be induced by the dynamical system (F, ψ, d). To establish chaos for the abstract similarity dynamics, we assume in this section the *separation condition*. Define the distance between two nonempty bounded sets A and B in $\mathcal{F}$ by $d(A, B) = \inf\{d(\mathbf{x}, \mathbf{y}): \mathbf{x} \in A, \mathbf{y} \in B\}$. The abstract similarity set $\mathcal{F}$ satisfies the separation condition of degree n if there exist a positive number ε_0 and a natural number n such that for arbitrary $i_1 i_2 \ldots i_n$ one can find $j_1 j_2 \ldots j_n$ such that

$$d\left(\mathcal{F}_{i_1 i_2 \ldots i_n}, \mathcal{F}_{j_1 j_2 \ldots j_n}\right) \geqslant \varepsilon_0. \tag{3.7}$$

We call ε_0 the *separation constant*.

For Poincaré chaos, Poisson stable motion is utilized to distinguish the chaotic behavior instead of the periodic motions in Devaney and Li–Yorke types. The existence of infinitely many unpredictable Poisson stable trajectories that lie in a compact set meet all the requirements of chaos. Based on this, chaos can appear in the dynamics on the quasi-minimal set which is the closure of a Poisson stable trajectory. Therefore the Poincarè chaos is referred to as the dynamics on the quasi-minimal set of trajectory initiated from unpredictable point. The next theorem shows that the Poincaré chaos is valid for the similarity dynamics.

Let us first give the definition of the unpredictability in the sense of the dynamics $(\mathcal{F}, \varphi, d)$.

Definition 3.1.1. [9] *A point $\mathcal{F}_{i_1 i_2 \ldots}$ from the set $\mathcal{F}$ is unpredictable if there exist a positive number ε_0 and sequences t_m, s_m, of natural numbers both of which diverge to infinity such that $\varphi^{t_m}(\mathcal{F}_{i_1 i_2 \ldots i_k \ldots}) \to \mathcal{F}_{i_1 i_2 \ldots i_k \ldots}$ as $m \to \infty$ and $d(\varphi^{t_m + s_m}(\mathcal{F}_{i_1 i_2 \ldots i_k \ldots}), \varphi^{t_m}(\mathcal{F}_{i_1 i_2 \ldots i_k \ldots})) \geqslant \varepsilon_0$ for each natural number m.*

The following theorem is an immediate consequence of the last definition.

Theorem 3.1.1. [7] *If the diameter and separation conditions are valid, then the similarity map possesses Poincaré chaos.*

Proof. Let us fix an unpredictable point of the dynamics as a member $\mathcal{F}_{i_1 i_2 \ldots i_k \ldots}$ of $\mathcal{F}$, which index sequence $i_1 i_2 \ldots i_k \ldots$ consists of subsequently chosen strings all of length one, then all possible strings of length two, etc, to infinity. The sample of this construction is seen in example 2.1.4 for the two symbolic alphabets case. It is easily seen from the diameter condition that the point is Poisson stable. That is, there exists a sequence of integers, t_m, divergent to infinity such that $\varphi^{t_m}(\mathcal{F}_{i_1 i_2 \ldots i_k \ldots}) \to \mathcal{F}_{i_1 i_2 \ldots i_k \ldots}$ as $n \to \infty$. Consider the unpredictability property. We will show that there is a sequence s_m divergent to infinity such that $d(\varphi^{t_m + s_m}(\mathcal{F}_{i_1 i_2 \ldots i_k \ldots}), \varphi^{t_m}(\mathcal{F}_{i_1 i_2 \ldots i_k \ldots})) \geqslant \varepsilon_0$, where ε_0 is the separation constant for the space $(\mathcal{F}, d)$. Assume, on the contrary, that there is no such sequence. Then, $d(\varphi^{t_m + i}(\mathcal{F}_{i_1 i_2 \ldots i_k \ldots}), \varphi^{t_m}(\mathcal{F}_{i_1 i_2 \ldots i_k \ldots})) < \varepsilon_0$, for all natural i. This contradicts the diameter condition and the choice of the initial point, since this means the absence of the initial string of length n in the index sequence of the point $\varphi^{t_m}(\mathcal{F}_{i_1 i_2 \ldots i_k \ldots})$. □

Now the following assertion is valid.

Theorem 3.1.2. [7] *If the diameter, separation and image conditions hold, then the dynamics (F, ψ, d) possesses Poincaré chaos.*

In the following theorem, we prove that the similarity map φ possesses the three ingredients of Devaney chaos, namely density of periodic points, transitivity and sensitivity. A point $\mathcal{F}_{i_1 i_2 i_3 \ldots}$ of $\mathcal{F}$ is periodic with period n if its index consists of endless repetitions of a block of n terms.

Theorem 3.1.3. [7] *If the diameter and separation conditions hold, then the similarity map is chaotic in the sense of Devaney.*

Proof. Fix a member $\mathcal{F}_{i_1 i_2 \dots i_n \dots}$ of $\mathcal{F}$ and a positive number ε. Find a natural number k such that $\mathrm{diam}(\mathcal{F}_{i_1 i_2 \dots i_k}) < \varepsilon$ and choose a k-periodic element $\mathcal{F}_{i_1 i_2 \dots i_k i_1 i_2 \dots i_k \dots}$ of $\mathcal{F}_{i_1 i_2 \dots i_k}$. It is clear that the periodic point is an ε-approximation for the considered member. The density of periodic points is thus proved.

Next, utilizing the diameter condition, the transitivity will be proved if we show the existence of an element $\mathcal{F}_{i_1 i_2 \dots i_n \dots}$ of $\mathcal{F}$ such that for any subset $\mathcal{F}_{i_1 i_2 \dots i_k}$ there exists a sufficiently large integer p so that $\varphi^p(\mathcal{F}_{i_1 i_2 \dots i_n \dots}) \in \mathcal{F}_{i_1 i_2 \dots i_k}$. This is true since we can construct the sequence $i_1 i_2 \dots i_n \dots$ such that it contains all sequences of the type $i_1 i_2 \dots i_k$ as blocks. For sensitivity, fix a point $\mathcal{F}_{i_1 i_2 \dots}$ in $\mathcal{F}$ and an arbitrary positive number ε. Due to the diameter condition there exist an integer k and element $\mathcal{F}_{i_1 i_2 \dots i_k j_{k+1} j_{k+2} \dots} \neq \mathcal{F}_{i_1 i_2 \dots i_k i_{k+1} i_{k+2} \dots}$ such that $d\left(\mathcal{F}_{i_1 i_2 \dots i_k i_{k+1} \dots}, \mathcal{F}_{i_1 i_2 \dots i_k j_{k+1} j_{k+2} \dots}\right) < \varepsilon$. We choose $j_{k+1}, j_{k+2}, \dots$ such that $d\left(\mathcal{F}_{i_{k+1} i_{k+2} \dots i_{k+n}}, \mathcal{F}_{j_{k+1} j_{k+2} \dots j_{k+n}}\right) > \varepsilon_0$, by the separation condition. This proves the sensitivity. □

From the last assertion the following theorem is implied easily.

Theorem 3.1.4. [7] *If the diameter, separation and image conditions hold, then the dynamical system* (F, ψ, d) *is Devaney chaotic.*

In addition to the Devaney and Poincaré chaos, it can be shown that the Li–Yorke chaos also takes place in the dynamics of the map φ. The proof of the following theorem is similar to that of theorem 6.35 in [24] for the shift map defined on the space of symbolic sequences.

Theorem 3.1.5. [7] *The similarity map is Li–Yorke chaotic if the diameter and separation conditions hold.*

The last assertion easily implies the next result.

Theorem 3.1.6. [7] *If the diameter, separation and image conditions hold, then the dynamics* (F, ψ, d) *possesses Li–Yorke chaos.*

We shall call the chaos for dynamics $(\mathcal{F}, \phi, d)$ *abstract similarity chaos.* Thus if the diameter and separation conditions are valid then the dynamics is with the abstract similarity chaos. Correspondingly, the dynamics discovered for (F, ψ, d) by the labeling is said to be *domain structured chaos.* Obviously, the domain structured chaos means not only the presence of the three fundamental types of chaos, but also an invitation for further research on the basis of the method.

Example 3.1.2. [7] *We have shown that the space of symbolic strings is an abstract similarity set in example* 3.1.1. *One can see that* $\mathcal{S} = \mathcal{S}_0 \cup \mathcal{S}_1$, *for the set where* $\mathcal{S}_0 = \{\mathcal{S}_{0 s_2 s_3 \dots}\}$ *and* $\mathcal{S}_1 = \{\mathcal{S}_{1 s_2 s_3 \dots}\}$, *hence*

$$
\begin{aligned}
d(\mathcal{S}_0, \mathcal{S}_1) &= \inf\{d(s, t): s \in \mathcal{S}_0, t \in \mathcal{S}_1\} \\
&= d(000\ldots, 1000\ldots) \\
&= d(0111\ldots, 111\ldots) = 1.
\end{aligned}
$$

Therefore, the separation condition of degree 1 *holds with the separation constant* ε_0 *equal to unity. According to the results of this section, abstract similarity dynamics and, consequently, the Bernoulli shift is chaotic in the sense of Poincaré, Li–Yorke and Devaney. That is, we confirm one more time the presence of chaos, which has been proven for the dynamics in* [8, 9, 26, 61].

In the last example the idea of the domain structured dynamics application is shown very clearly. Now consider the problem generally. Let $\mathcal{F}$ be an abstract label set for a set F. Assume that the separation and diameter conditions are valid and by theorems 3.1.1, 3.1.3 and 3.1.5 the abstract dynamics $(\mathcal{F}, \varphi, d)$ has abstract similarity chaos, i.e. it is Poincaré, Li–Yorke and Devaney chaotic. Consider a dynamical system (F, ψ, d). Assume that the mappings ψ and φ agree such that if $\mathcal{F}_{i_1 i_2 \ldots}$ is a label for an element f of F, then $\phi(f) = \mathcal{F}_{i_2 i_3 \ldots}$ is a label for $\psi(f)$. One can easily see that if the agreement is valid then the dynamics (F, ψ, d) is also Poincaré, Li–Yorke and Devaney chaotic. That is, theorems 3.1.8, 3.1.5 and 3.1.10 are valid. To be brief, we shall say that the dynamics has *domain structured chaos*. That is, we have proved in the last example that the symbolic dynamics is domain structured chaotic. Assume also the case that a metric space (F, d) is not provided with a dynamics. Then we say that the metric space is assigned with domain structured chaos, if there exists an abstract label set $\mathcal{F}$ for F such that $(\mathcal{F}, \varphi, d)$ is abstract similarly chaotic. The last notion is useful for the fractal analysis of chapter 4. Another important feature of the abstract similarity chaos and domain structured chaos is that all label points and labeled points are members of the chaos domain. In the case of Poincaré chaos they are points of a quasi-minimal set. Let us consider more examples.

3.1.5 Examples

Example 3.1.3. *Consider the finite metric space* (F, d) *of two elements f and g. Assign to f and g infinitely many labels* $\mathcal{F}_{1 i_2 \ldots}$ *and* $\mathcal{F}_{2 i_2 \ldots}$, *respectively, such that* $i_k = 1, 2$ *for all* $k = 2, 3, \ldots$. *It is clear that the separation and diameter conditions are valid for the dynamics* $(\mathcal{F}, \phi, d)$. *Now introduce in the space* (F, d) *the mapping* ψ, *which agrees with* φ. *Thus we have obtained domain structured chaotic dynamics* (F, ψ, d). *The dynamics will be useful to discuss random processes.*

Example 3.1.4. *Consider the Cantor middle set,* $\mathcal{C}$, *in the section* [0, 1]. *By the known procedure* [26, 61] *assign to each point of the set the label* $\tau_{i_1 i_2 \ldots}$, $i_k = 0, 1$, $k = 1, 2, \ldots$, *and denote by* $\mathcal{T}$ *the set of labels. Specifically, rectangles* $\mathcal{T}_1$ *and* $\mathcal{T}_2$ *are for points of the Cantor set in sections* [0, 1/3] *and* [2/3, 1], *respectively, rectangles* $\mathcal{T}_{11}$ *and* $\mathcal{T}_{12}$ *are for points of the Cantor set in sections* [0, 1/9] *and* [2/9, 1/3], *respectively, etc. Introduce the distance*

$$d(\tau_{i_1 i_2 \ldots}, \tau_{j_1 j_2 \ldots}) = |c_1 - c_2|,$$

where $|\cdot|$ *is the absolute value, and* $\tau_{i_1 i_2 \ldots}$ *with* $\tau_{j_1 j_2 \ldots}$ *are labels for points* c_1 *and* c_2 *in the Cantor set. The labeling procedure confirms the abstract similarity chaos and domain structured chaos for the Cantor set. Thus we have obtained the chaos for the geometrical object, without topological equivalence of the logistic map and the symbolic dynamics* [26].

Let us consider another version of the discussion. One can apply the distance

$$d(\tau_{i_1 i_2 \ldots}, \tau_{j_1 j_2 \ldots}) = \sum_{k=1}^{\infty} \frac{|i_k - j_k|}{2^{k-1}}$$

for the label set. Then repeat the analysis with the distance $d(c_1, c_2) = d(\tau_{i_1 i_2 \ldots}, \tau_{j_1 j_2 \ldots})$ *in* $\mathcal{C}$ *and additionally note the equivalence of the two metrics,*

$$1/3|c_1 - c_2| \leqslant \sum_{k=1}^{\infty} \frac{|i_k - j_k|}{2^{k-1}} \leqslant 9/2|c_1 - c_2|,$$

to make a conclusion on the domain structured chaos on $\mathcal{C}$.

Example 3.1.5. *(multi-folded scalar baker transformation) Consider the baker map, which is equal to* $B(x) = nx - i$ *on* $(i/n, (i+1)/n]$, $i = 0, 1, \ldots, n-1$. *The graph of the map is seen in figure* 3.1. *Denote* $\mathcal{F}_i = (i/n, (i+1)/n]$, $i = 0, 1, \ldots, n-1$, *the first order rectangles. Next find that* $\mathcal{F}_j = \cup_{k=0}^{n-1} \mathcal{F}_{jk}$, $j = 0, 1 \ldots, n-1$, *where* $B(\mathcal{F}_{jk}) = \mathcal{F}_j$, $k = 0, 1, \ldots, n-1$. *Continue the process, considering inductively* $\mathcal{F}_{i_1 i_2 \ldots i_{n-1} i_n} = \cup_{k=0}^{n-1} \mathcal{F}_{i_1 i_2 \ldots i_n k}$, *such that* $B(\mathcal{F}_{i_1 i_2 \ldots i_n k}) = \mathcal{F}_{i_1 i_2 \ldots i_{n-1} k}$, $k = 0, 1, \ldots, n-1$, *if all rectangles* $\mathcal{F}_{i_1 i_2 \ldots i_{n-1} i_n}$ *for a fixed n are determined. The label set* $\mathcal{F}$ *is the collection of points* $\mathcal{F}_{i_1 i_2 \ldots}$ *such that a point with fixed* $i_1 i_2 \ldots$ *belongs to* $\mathcal{F}_{i_1 i_2 \ldots i_n}$ *with all natural n. The reader can easily verify that the set is an abstract similarity set. The separation condition is valid for sets* $\mathcal{F}_{ij}$, $i, j = 0, 1$, *with* $\varepsilon_0 = 1/(2n)$. *Moreover, if* $n > 2$, *then the*

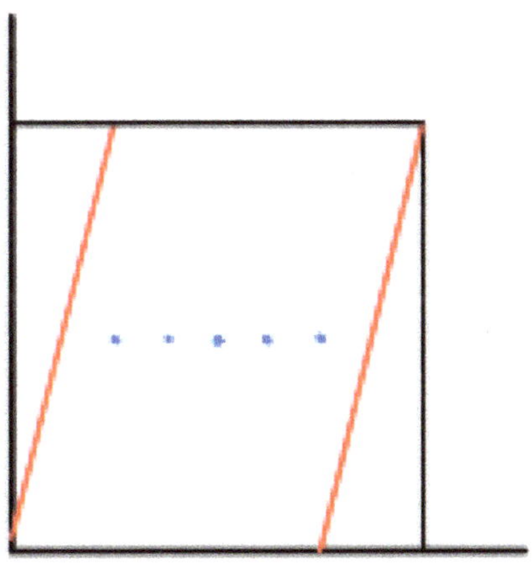

Figure 3.1. The graph of the multi-folded baker map B.

separation condition is of the first degree with $\varepsilon_0 = 1/n$. *The final and essential remark is that the map B and the similarity map* φ *agree at each point of the domain such that* $B(x) = \varphi(\mathcal{F}_{i_1 i_2 \ldots})$ *if the number x in* [0, 1] *is labeled with* $\mathcal{F}_{i_1 i_2 \ldots}$. *Thus the domain structured chaos is approved for the baker transformation. It is valuable that the domain of chaos is the unit section and is not a fractal set. We have proved that the unit section is a quasi-minimal set for the baker map. Our results shed more light on the important property of the recurrence of the map, since this time it is combined with the unpredictability property. Thus we should recognize that the classical concept of ergodicity is not sufficient for the description of the process* [47] *and should be developed with new features.*

3.2 Hyperbolic chaos

The abstract hyperbolic sets are introduced in this section. They are developments of abstract similarity sets. The abstract similarity map is applied to the sets. Continuous and differentiable mappings as well as the rate of convergence and transversal manifolds are not under discussion. Our suggestions are for more neat comprehension of chaos in the domains. The novelties can serve for revisited models as well as motivate dynamical research of new models.

Hyperbolic sets were introduced by Smale in the famous papers [56, 57]. They are fundamental for chaos theory, since they describe domains where the dynamics occurs in profound detail. The genius of Poincaré who explored the complexity in celestial mechanics [53] and the scrupulous investigation of the Van der Pol oscillator by Cartright, Littlewood and Levinson [23, 40] enabled the insights of Smale. An infinite number of hyperbolic points in a bounded area, symbolic dynamics, manifolds, Markov partitions and differentiable mappings form the background of the theory [3, 16–21, 26, 45, 49, 55, 61].

Recently, we have introduced the concept of domain structured chaos [4, 6, 7], which may become a useful tool for the analysis of complex dynamics [7]. This time the new concept of *abstract hyperbolic sets* is considered to continue the line of research. What is suggested is not the application of the symbolic dynamics through topological equivalence, which requires additional advanced conditions for the establishment of the relation [26, 33, 45]. Moreover, we do not utilize finite partitioning [17, 19] and even countable [21] partitioning for the involvement of the symbolic dynamics to explore the complexity. We perform the complete indexing of the domain, in other words infinitesimal uncountable partitioning and determining the map, the orbits of which are constructed through shifting in indexes such that the research can be focused on chaotic behavior. By utilizing the *separation* and *diameter conditions* [6, 7], we explore the *domain structured chaos* in the abstract hyperbolic sets. The discussion initiated in this paper can provide a precise description of chaotic dynamics as well as hyperbolic sets, explore relations which either have not been seen in former studies or can make more sense of them. For example, in the current chapter, a new proof of the Smale horseshoe and two-dimensional baker transformation chaos is provided, which is free of the equivalence application. The results are complemented by Poincaré type chaos and isolated dynamics on unstable sets.

In [49] Ornstein proved that stochastic processes such as Bernoulli and Markov shifts, Anosov flows and Sinai's billiards, ergodic automorphisms of the n-torus, and the continued fraction transform are in fact isomorphic. This makes us confident that the current suggestions can be useful for the further investigation of chaos in various hyperbolic structures.

3.2.1 Hyperbolic similarity dynamics

In this subsection we introduce a formal set and a map, which generalize the abstract similar dynamics considered in the last section. Another suggestion involves the abstraction of the concept of hyperbolic sets [17, 57], which is understood, in our research, as renouncing some of the former properties and considering new ones, which are obvious and useful.

Let a space $(\mathcal{F}, d)$ be given with metric d. The elements of the set are labels such that

$$\mathcal{F} = \{\mathcal{F}_{\ldots i_{-2} i_{-1} i_0 \cdot i_1 i_2 \ldots} : i_k = 1, 2, \ldots, m, \;\; k = \ldots -2, -1, 0, 1, 2, \ldots\}, \tag{3.8}$$

where m is a fixed natural number and the symbol $\cdot$ is used to determine the dynamics of the present research.

The set $\mathcal{F}$ is called the *abstract hyperbolic set* and its elements are called *abstract hyperbolic points*. The set of points

$$\mathcal{F}_{i_1 i_2 \ldots} = \left\{\mathcal{F}_{\ldots j_{-2} j_{-1} j_0 \cdot i_1 i_2 \ldots i_n \ldots} : j_k = 1, 2, \ldots, m, \;\; k = \ldots -2, -1, 0\right\}, \tag{3.9}$$

is the *stable set* for a fixed point $\mathcal{F}_{\ldots i_{-k} \ldots i_{-2} i_{-1} i_0 \cdot i_1 i_2 \ldots i_n \ldots}$, and the collection

$$\mathcal{F}_{\ldots i_{-k} \ldots i_{-2} i_{-1} i_0} = \left\{\mathcal{F}_{\ldots i_{-2} i_{-1} i_0 \cdot j_1 j_2 \ldots} : j_k = 1, 2, \ldots, m, \;\; k = 1, 2, \ldots\right\} \tag{3.10}$$

is said to be the *unstable set* of the point.

The following unions, rectangles, are needed in what follows,

$$\mathcal{F}_{i_{-k} i_{-k+1} \ldots i_0 \cdot i_1 i_2 \ldots i_n} = \bigcup_{j_k = 1, 2, \ldots, m} \mathcal{F}_{\ldots i_{-k-1} i_{-k} i_{-k+1} \ldots i_0 \cdot i_1 i_2 \ldots i_n i_{n+1} \ldots}, \tag{3.11}$$

where indices $i_{-k}, i_{-k+1}, \ldots, i_0, i_1, i_2, \ldots, i_n$ are fixed,

$$\mathcal{F}_{i_1 i_2 \ldots i_n} = \bigcup_{j_k = 1, 2, \ldots, m} \mathcal{F}_{\ldots i_{-k-1} i_{-k} i_{-k+1} \ldots i_0 \cdot i_1 i_2 \ldots i_n i_{n+1} \ldots}, \tag{3.12}$$

where indices $i_1, i_2, \ldots, i_n$ are fixed and

$$\mathcal{F}_{i_{-k} i_{-k+1} \ldots i_0} = \bigcup_{j_k = 1, 2, \ldots, m} \mathcal{F}_{\ldots i_{-k-1} i_{-k} i_{-k+1} \ldots i_0 \cdot i_1 i_2 \ldots i_n i_{n+1} \ldots}, \tag{3.13}$$

where indices $i_{-k}, i_{-k+1}, \ldots, i_0$ are fixed.

We will say that the *diameter condition* is valid for the sets if for every fixed sequence $\ldots i_{-k} i_{-k+1} \ldots i_0 \cdot i_1 i_2 \ldots i_n \ldots$ it is true that

$$\sup_{i_{-k}i_{-k+1}\ldots i_0 \cdot i_1 i_2 \ldots i_n} \operatorname{diam} \mathcal{F}_{i_{-k}i_{-k+1}\ldots i_0 \cdot i_1 i_2 \ldots i_n} \to 0 \text{ as } k, n \to \infty, \tag{3.14}$$

where $\operatorname{diam}(A) = \sup\{d(\mathbf{x}, \mathbf{y}) \colon \mathbf{x}, \mathbf{y} \in A\}$ for a subset A of $\mathcal{F}$.

Determine the following one-to-one and onto map $\varphi \colon \mathcal{F} \to \mathcal{F}$, such that

$$\varphi(\mathcal{F}_{\ldots i_{-k}\ldots i_{-2}i_{-1}i_0 \cdot i_1 i_2 \ldots i_n \ldots}) = \mathcal{F}_{\ldots i_{-k}\ldots i_{-2}i_{-1}i_0 i_1 \cdot i_2 \ldots i_n \ldots}.$$

Let us fix a point $\mathcal{F}_{\ldots i_{-k}\ldots i_{-2}i_{-1}i_0 \cdot i_1 i_2 \ldots i_n \ldots}$ of the hyperbolic set and an arbitrary element, $\mathcal{F}_{\ldots j_{-2}j_{-1}j_0 \cdot i_1 i_2 \ldots i_n \ldots}$, from the stable manifold of the point. From the definition of the stable set and the diameter property it follows that

$$d(\varphi^n(\mathcal{F}_{\ldots j_{-2}j_{-1}j_0 \cdot i_1 i_2 \ldots i_n \ldots}), \varphi^n(\mathcal{F}_{\ldots i_{-k}\ldots i_{-2}i_{-1}i_0 \cdot i_1 i_2 \ldots i_n \ldots})) \to 0$$

as $n \to \infty$.

Similarly, it is true that

$$d(\varphi^n(\mathcal{F}_{\ldots i_{-2}i_{-1}i_0 \cdot j_1 j_2 \ldots j_n \ldots}), \varphi^n(\mathcal{F}_{\ldots i_{-k}\ldots i_{-2}i_{-1}i_0 \cdot i_1 i_2 \ldots i_n \ldots})) \to 0$$

as $n \to -\infty$ for points of the unstable manifold.

It is clear that

$$\mathcal{F} \supseteq \mathcal{F}_{i_1} \supseteq \mathcal{F}_{i_1 i_2} \supseteq \ldots \supseteq \mathcal{F}_{i_1 i_2 \ldots i_n} \supseteq \mathcal{F}_{i_1 i_2 \ldots i_n i_{n+1}} \ldots$$

and

$$\mathcal{F} \supseteq \mathcal{F}_{i_0} \supseteq \mathcal{F}_{i_{-1}i_0} \supseteq \ldots \supseteq \mathcal{F}_{i_{-n}i_{-n+1}\ldots i_0} \supseteq \mathcal{F}_{i_{-n-1}i_{-n}\ldots i_0} \ldots.$$

Considering iterations of the map, one can verify that

$$\varphi^n(\mathcal{F}_{i_1 i_2 \ldots i_n}) = \mathcal{F} \tag{3.15}$$

and

$$\varphi^{-n}(\mathcal{F}_{i_{-n}i_{-n+1}\ldots i_0}) = \mathcal{F} \tag{3.16}$$

for arbitrary natural number n. The relations (3.15) and (3.16) give us a reason to call φ a *similarity map* and the number n the *degree of similarity*. We called it the *abstract similarity map* in [7].

3.2.2 Abstract similarity chaos

Denote by $d(A, B) = \inf\{d(\mathbf{x}, \mathbf{y}) \colon \mathbf{x} \in A, \mathbf{y} \in B\}$ the function of two bounded subsets A and B of the abstract hyperbolic set $\mathcal{F}$. The space $(\mathcal{F}, d)$ satisfies the *positive separation condition* of degree n if there exist a positive number ε_0 and a natural number n such that for arbitrary indices $i_1 i_2 \ldots i_n$ one can find indices $j_1 j_2 \ldots j_n$ such that

$$d\left(\mathcal{F}_{i_1 i_2 \ldots i_n}, \mathcal{F}_{j_1 j_2 \ldots j_n}\right) \geqslant \varepsilon_0. \tag{3.17}$$

Similarly, the *negative separation condition* of degree n is valid for the space $(\mathcal{F}, d)$ if there exist a positive number ε_0 and a natural number n such that for arbitrary indices $i_{-n+1}i_{-n+2}\ldots i_0$ one can find indices $j_{-n+1}j_{-n+2}\ldots j_0$ such that

$$d\left(\mathcal{F}_{i_{-n+1}i_{-n+2}\dots i_0}, \mathcal{F}_{j_{-n+1}j_{-n+2}\cdots j_0}\right) \geqslant \varepsilon_0. \tag{3.18}$$

We say that the separation condition is valid if both positive and negative separation conditions are fulfilled.

A point $\mathcal{F}_{\dots i_{-k}\dots i_{-2}i_{-1}i_0 \cdot i_1 i_2 \dots i_n \dots}$ from $\mathcal{F}$ is periodic with period n if its index consists of endless repetitions of a block of n terms.

Theorem 3.2.1. *If the diameter and positive separation conditions are valid, then the similarity map φ is chaotic in the sense of Devaney.*

Proof. The transitivity will be proved if there exists an element $\mathcal{F}_{\dots i_{-k}\dots i_{-2}i_{-1}i_0 \cdot i_1 i_2 \dots i_n \dots}$ of $\mathcal{F}$ such that for any finite sequence $j_{-k}\cdots j_{-2}j_{-1}j_0 \cdot j_1 j_2 \cdots j_n$ one can find that for a sufficiently large p, the point $\varphi^p(\mathcal{F}_{\dots i_{-k}\dots i_{-2}i_{-1}i_0 \cdot i_1 i_2 \dots i_n \dots})$ belongs to $\mathcal{F}_{j_{-k}\cdots j_{-2}j_{-1}j_0 \cdot j_1 j_2 \cdots j_n}$. This is true, since one can construct a sequence $\dots i_1 i_2 \dots i_n \dots$, which contains all the sequences of the type $j_{-k}\cdots j_{-2}j_{-1}j_0 \cdot j_1 j_2 \cdots j_n$ as blocks [10, 61].

Fix a member $\mathcal{F}_{\dots i_{-k}\dots i_{-2}i_{-1}i_0 \cdot i_1 i_2 \dots i_n \dots}$ of the label set and a positive number δ. Find a sufficiently large number k such that $\mathrm{diam}(\mathcal{F}_{i_{-k}i_{-k+1}\dots i_0 \cdot i_1 i_2 \dots i_k}) < \delta$ and choose a $2k$-periodic element $\mathcal{F}_{\dots i_{-k}i_{-k+1}\dots i_0 \cdot i_1 i_2 \dots i_k \dots}$ of $\mathcal{F}_{i_{-k}i_{-k+1}\dots i_0 \cdot i_1 i_2 \dots i_k}$. It is clear that the periodic point is an δ-approximation for the considered member. Thus the density of periodic points is proved.

Fix a point $\mathcal{F}_{\dots i_{-k}\dots i_{-2}i_{-1}i_0 \cdot i_1 i_2 \dots i_n \dots}$ of the label set and an arbitrary positive number ε. By the diameter condition there exist an integer k and point $\mathcal{F}_{\dots i_{-k}\dots i_{-2}i_{-1}i_0 \cdot i_1 i_2 \dots i_k j_{k+1} j_{k+2} \cdots} \neq \mathcal{F}_{\dots i_{-k}\dots i_{-2}i_{-1}i_0 \cdot i_1 i_2 \dots i_n \dots}$ such that $d(\mathcal{F}_{\dots i_{-k}\dots i_{-2}i_{-1}i_0 \cdot i_1 i_2 \dots i_n \dots}, \mathcal{F}_{\dots i_{-k}\dots i_{-2}i_{-1}i_0 \cdot i_1 i_2 \dots i_k j_{k+1} j_{k+2} \cdots}) < \varepsilon$. We choose $j_{k+1}, j_{k+2}, \dots$ such that $d(\mathcal{F}_{i_{k+1}i_{k+2}\dots i_{k+n}}, \mathcal{F}_{j_{k+1}j_{k+2}\cdots j_{k+n}}) > \varepsilon_0$ by the separation condition. This proves the sensitivity. □

In [9] Poisson stable motion is utilized to distinguish chaotic behavior from periodic motions in the Devaney and Li–Yorke types. The dynamics is called Poincaré chaos.

Theorem 3.2.2. *The similarity map φ is chaotic in the Poincaré sense, if the diameter and positive separation conditions are valid.*

The proof of the last theorem is similar to the proof of theorem 3.1.3.

In addition to the Devaney and Poincaré chaos, it can be shown that the Li–Yorke chaos exists for the similarity map φ. The proof is similar to that of theorem 6.35 in [24]. Thus we have obtained that there is Poincaré, Li–Yorke and Devaney chaos on each unstable set of the similarity map.

Now we suggest considering a new concept for chaos analysis. Let us agree that domain structured chaos is valid for decreasing time if the ingredients of Poincaré, Li–Yorke and Devaney chaos are correct for an abstract hyperbolic set with the inverse map ϕ^{-1}. The following theorems are valid.

Theorem 3.2.3. *If the diameter and negative separation conditions are valid, then the inverse similarity map φ^{-1} is chaotic in the sense of Devaney.*

Theorem 3.2.4. *The inverse similarity map φ^{-1} is chaotic in the Poincaré sense, if the diameter and negative positive separation properties are valid.*

Theorem 3.2.5. *If the diameter and negative separation conditions are valid, then the inverse similarity map φ^{-1} is chaotic in the sense of Li–Yorke.*

Finally, if for the space $(\mathcal{F}, d)$ the diameter, positive separation and negative separation conditions are valid, then the dynamics $(\mathcal{F}, \varphi, d)$ admits the *abstract hyperbolic similarity chaos.*

3.2.3 Hyperbolic domain structured chaos

Consider a dynamics (F, ψ, d) with a state space F and invertable mapping ψ such that (F, d) is a compact metric space. Assume that an abstract hyperbolic set $\mathcal{F}$ is a label set for F and the labeling is unique. Keeping that $d(\mathcal{F}_{\ldots i_{-k}\ldots i_{-2}i_{-1}i_0 \cdot i_1 i_2 \ldots i_n \ldots}, \mathcal{F}_{\ldots j_{-2}j_{-1}j_0 \cdot j_1 j_2 \ldots j_n \ldots}) = d(f, g)$ for labels of elements f and g, we introduce the hyperbolic abstract similarity dynamics $(\mathcal{F}, \varphi, d)$. Thus the space $(\mathcal{F}, d)$ is also compact. In the case that the dynamics is chaotic and maps ψ and φ agree, we shall say that (F, ψ, d) has hyperbolic domain structured chaos. Let us recall that the maps agree if the label of an image is the value of the abstract similarity map on the label for the pre-image. We say that the *image condition* is valid if the mappings agree. The following assertions are true.

Theorem 3.2.6. *If the diameter, positive separation and image conditions are valid, then the dynamics (F, ψ, d) is Devaney chaotic.*

Theorem 3.2.7. *The dynamics (F, ψ, d) is chaotic in the Poincaré sense if the diameter, positive separation and image conditions are valid.*

Theorem 3.2.8. *If the diameter, positive separation and image conditions hold, then the inverse map ψ^{-1} is chaotic in the sense of Devaney.*

Theorem 3.2.9. *The map ψ^{-1} is chaotic in the Poincaré sense if the diameter, negative positive separation and image conditions are valid.*

Theorem 3.2.10. *If the diameter, negative separation and image conditions hold, then ψ^{-1} is Li–Yorke chaotic.*

3.2.4 Examples

The following analyses of popular models illustrate the strength of our suggestions.

Example 3.2.1. [10] *Let us take into account the following space of bi-infinite sequences* [61],

$$\Sigma^2 = \{s = (\ldots s_{-2}s_{-1}.\ s_0 s_1 s_2 \ldots) \colon s_j = 0 \text{ or } 1 \text{ for each } j\}$$

with the metric

$$d[s, \bar{s}] = \sum_{k=-\infty}^{\infty} \frac{|s_k - \bar{s}_k|}{2^{|k|}},$$

where $s = (\ldots s_{-2}s_{-1}.\ s_0s_1s_2\ldots)$, $\bar{s} = (\ldots \bar{s}_{-2}\bar{s}_{-1}.\ \bar{s}_0\bar{s}_1\bar{s}_2\ldots) \in \Sigma^2$. *The shift map* σ: $\Sigma^2 \to \Sigma^2$ *is defined as*

$$\sigma(\ldots s_{-2}s_{-1}.\ s_0s_1s_2\ldots) = (\ldots s_{-2}s_{-1}s_0.\ s_1s_2\ldots).$$

One can easily show that the dynamics (Σ^2, σ, d) *has hyperbolic domain structured chaos.*

Example 3.2.2. *(Smale horseshoe.) As an example consider the Smale horseshoe map* [57]. *To prove the dynamics in the set traditionally, a bi-infinite topologically equivalent symbolic dynamics has been used* [61]. *We will demonstrate the domain structured chaos for the result. Take the closed unit square in figure* 3.2*(a). Contract the square vertically with a coefficient k less than* 1/2 *and stretch it horizontally with a factor* 1/*k. Then bend it as a horseshoe-shaped figure and place the result on the square, as can be seen in figure* 3.2*(b). Denote by S the map and consider the Smale hyperbolic set, that is all bi-infinite trajectories in the square united. Let us start with vertical rectangles* $V_{.0}$ *and* $V_{.1}$, *which can be seen in figure* 3.3*(a). Their images are horizontal rectangles* $H_{0.}$ *and* $H_{1.}$ *such that* $S(V_{.j}) = H_{j.}$, $j = 0, 1$, *see figure* 3.3*(b). The rectangles are closed as well in what follows. We use indexes* $i_j = 0$ *or 1 with j as integers and* $\cdot$ *as a dot-index. One can determine recurrently the set of*

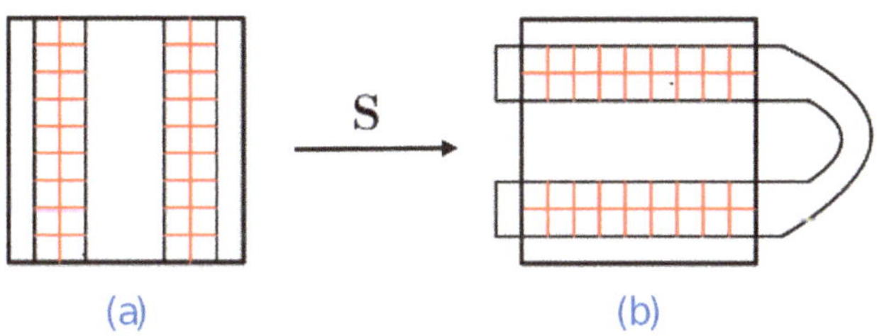

Figure 3.2. Design of a Smale horseshoe map.

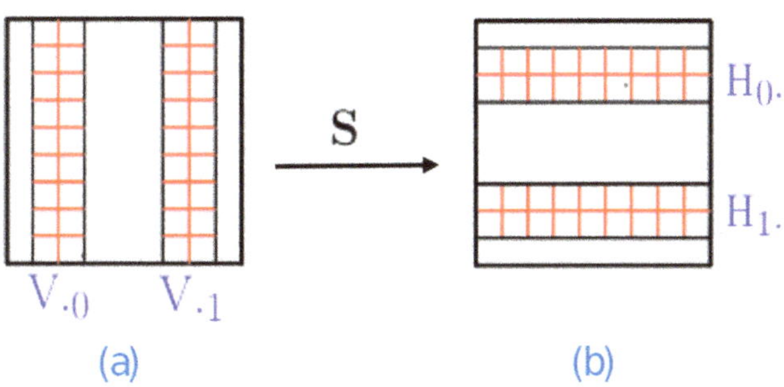

Figure 3.3. The first rectangles of the horseshoe construction.

horizontal rectangles $H_{i_{-n}i_{-n+1}\ldots i_{-1}i_0\cdot}$ *such that* $S^{-1}(H_{i_{-n}i_{-n}\ldots i_{-1}i_0\cdot}) = H_{i_{-n}i_{-n+1}\ldots i_{-1}\cdot}$ *and* $H_{i_{-n}\ldots i_{-1}i_0\cdot} \subset H_{i_{-n}i_{-n+1}\ldots i_{-1}\cdot}$. *Similarly, we determine the set of vertical rectangles* $V_{\cdot i_1 i_2\ldots i_{n-1}i_n}$ *such that* $S(V_{\cdot i_1 i_2\ldots i_n i_n}) = V_{\cdot i_2\ldots i_{n-1}i_n}$, $V_{\cdot i_1\ldots i_n} \subset V_{\cdot i_2\ldots i_n}$. *The first rectangles of the series are seen in figure* 3.3. *There exists a unique horizontal section* $h_{\ldots i_{-n}i_{-n+1}\ldots i_{-1}i_0\cdot}$, *the intersection of the infinite rectangles included,* $H_{i_0\cdot}$, $H_{i_{-1}i_0\cdot}$, $H_{i_{-2}i_{-1}i_0\cdot}$, …. *Similarly, one can determine a vertical line* $v_{\cdot i_1 i_2\ldots i_n i_{n+1}\ldots}$ *as the intersection of the infinite sequence of the vertical rectangles* $V_{\cdot i_1}$, $V_{\cdot i_1 i_2}$, $V_{\cdot i_1 i_2 i_3}$…. *Denote by* $\mathcal{F}_{\ldots i_{-n}i_{-n+1}\ldots i_{-1}i_0\cdot i_1 i_2 i_3\ldots}$ *the intersection of the vertical line* $v_{\cdot i_1 i_2\ldots i_n i_{n+1}\ldots}$ *and the horizontal line* $h_{\ldots i_{-n}i_{-n+1}\ldots i_{-1}i_0\cdot}$. *The point belongs to the Smale set. The diameter condition is correct, since the rectangle* $\mathcal{F}_{i_{-n}i_{-n+1}\ldots i_{-1}i_0\cdot i_1 i_2 i_3\ldots i_n}$ *is included in the intersection of the rectangles* $H_{i_{-n}i_{-n+1}\ldots i_{-1}i_0\cdot}$ *and* $V_{\cdot i_1 i_2\ldots i_{n-1}i_n}$. *The separation condition is correct for sets* $\mathcal{F}_{i\cdot j}$, $i, j = 0, 1$. *Thus the hyperbolic abstract similarity chaos is valid for the label set with the abstract similarity map* $\phi: \mathcal{F} \to \mathcal{F}$. *Finally, by using the method of rectangle construction, one can observe that* $S(x) = y$ *if and only if* x *and* y *are labeled* $\mathcal{F}_{\ldots i_{-2}i_{-1}i_0\cdot i_1 i_2 i_3\ldots}$ *and* $\mathcal{F}_{\ldots i_{-1}i_0 i_1\cdot i_2 i_3\ldots}$, *respectively. Indeed,* $\phi(\mathcal{F}_{\ldots i_{-2}i_{-1}i_0\cdot i_1 i_2 i_3\ldots}) = \mathcal{F}_{\ldots i_{-2}i_{-1}i_0 i_1\cdot i_2 i_3\ldots}$. *However,* $\mathcal{F}_{\ldots i_{-2}i_{-1}i_0\cdot i_1 i_2 i_3\ldots}$ *is an intersection of lines* $v_{\cdot i_1 i_2\ldots}$ *and* $h_{\ldots i_{-1}i_0\cdot}$, *and* $\mathcal{F}_{\ldots i_{-n}i_{-n+1}\ldots i_{-1}i_0 i_1\cdot i_2 i_3\ldots}$ *is the intersection of lines* $v_{\cdot i_2\ldots}$ *and* $h_{\ldots i_{-1}i_0 i_1\cdot}$. *Moreover,* $S^{-1}(h_{\ldots i_{-1}i_0\cdot}) = h_{\ldots i_{-2}i_{-1}\cdot}$ *and* $S(v_{\cdot i_1 i_2\ldots}) = v_{\cdot i_2 i_3\ldots}$. *That is,* $\phi(\mathcal{F}_{\ldots i_{-n}i_{-n+1}\ldots i_{-1}i_0\cdot i_1 i_2 i_3\ldots}) = \mathcal{F}_{\ldots i_{-n}i_{-n+1}\ldots i_{-1}i_0 i_1\cdot i_2 i_3\ldots}$. *Hence, the Smale horseshoe map has hyperbolic domain structured chaos, i.e. there is Poincaré, Li–Yorke and Devaney chaos for the map and its inverse.*

Example 3.2.3. *(the two-dimensional baker transformation.) Consider the baker map of the unit square in figure* 3.4*(a), which can also be called the two-dimensional baker's transformation. The original figure is stretched to that in figure* 3.4*(b), then cut vertically and the pieces are joined for the result of the mapping, the square in figure* 3.4*(c).*

Considering the notations of the horizontal rectangles and B *as the map, one can see that* $B^{-1}(H_{00\cdot}) = B^{-1}(H_{01\cdot}) = H_0\cdot$ *and* $B^{-1}(H_{10\cdot}) = B^{-1}(H_{11\cdot}) = H_1\cdot$, *where the center-dot* $\cdot$ *is applied to the hyperbolic indications.*

It is now clear that proceeding in this way we obtain the set of horizontal rectangles $H_{i_{-n}i_{-n+1}\ldots i_{-1}i_0\cdot}$, $n = 1, 2, \ldots$, $i_j = 0, 1$, *such that* $B^{-1}(H_{i_{-n}i_{-n+1}\ldots i_{-1}i_0\cdot}) = H_{i_{-n}i_{-n+1}\ldots i_{-1}\cdot}$ *for an arbitrary rectangle, and* $H_{i_{-n}i_{-n+1}\ldots i_{-1}i_0\cdot} \subset H_{i_{-n}i_{-n+1}\ldots i_{-1}\cdot}$. *That is, the sequence* $H_{i_0\cdot}$, $H_{i_{-1}i_0\cdot}$, … *is nested. Denote by* $h_{\ldots i_{-n}i_{-n+1}\ldots i_{-1}i_0\cdot}$ *the horizontal line, which is in all of the rectangles.*

The baker map transforms the pre-image in figure 3.5*(a) to the original square in figure* 3.5*(b). The pre-image contains the vertical rectangles* $V_{\cdot 0}$ *and* $V_{\cdot 1}$ *such that* $B(V_{\cdot 0}) = H_{0\cdot}$ *and* $B(V_{\cdot 1}) = H_{1\cdot}$. *Moreover, one can see that for the rectangles in figure* 3.6 *it is true that* $B(V_{\cdot 00}) = B(V_{\cdot 10}) = V_{\cdot 0}$ *and* $B(V_{\cdot 01}) = B(V_{\cdot 11}) = V_{\cdot 1}$.

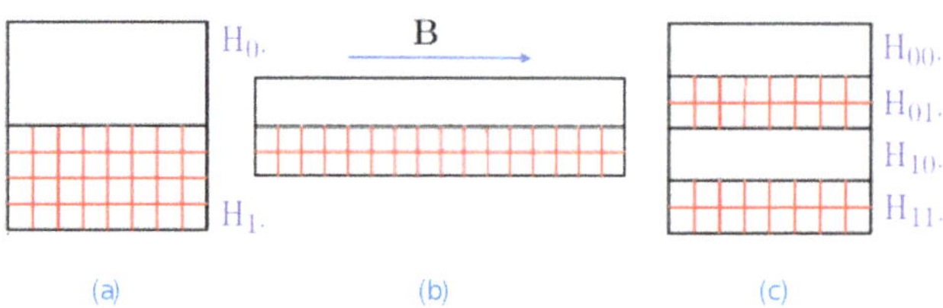

Figure 3.4. Design of the baker transformation.

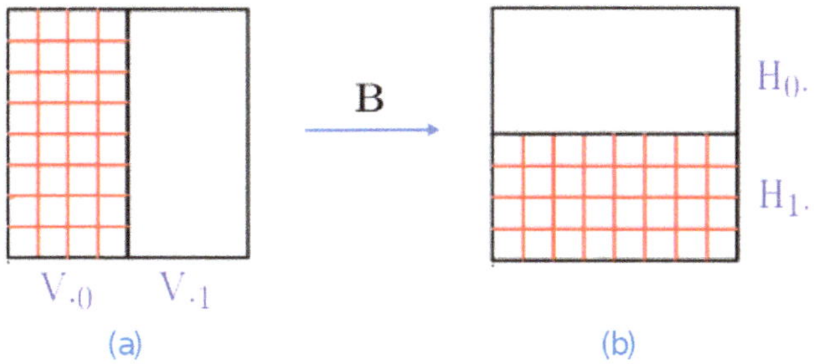

Figure 3.5. The first rectangles of the mapping.

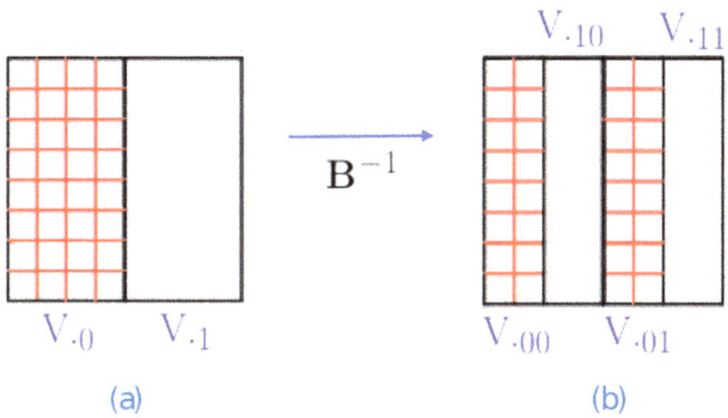

Figure 3.6. The second inverse iteration.

Generally, we have obtained a set of vertical rectangles $V_{\cdot i_0 i_1 \ldots i_n}$, $i_j = 0, 1$, *for arbitrary natural* n *such that* $B(V_{\cdot i_1 i_2 \ldots i_n}) = V_{\cdot i_2 \ldots i_n}$ *for an arbitrary rectangle and* $V_{\cdot i_1 i_2 \ldots i_n} \subset V_{\cdot i_2 \ldots i_n}$. *That is, the sequence* $V_{\cdot i_1}$, $V_{\cdot i_1 i_2}$, … *is nested. Denote by* $v_{\cdot i_1 i_2 \ldots i_n \ldots}$ *the vertical line, which is common for all of the nested vertical rectangles. We label by* $\mathcal{F}_{\ldots i_{-n} i_{-n+1} \ldots i_{-1} i_0 \cdot i_1 i_2 \ldots i_n \ldots}$ *the intersection of the horizontal line* $h_{\ldots i_{-n} i_{-n+1} \ldots i_{-1} i_0 \cdot}$, *and the vertical line* $v_{\cdot i_1 i_2 \ldots i_n \ldots}$ *and unite all the labels as the label set* $\mathcal{F}$. *It is true that* $B(x) = y$ *if and only if* x *and* y *are labeled with* $\mathcal{F}_{\ldots i_{-n} i_{-n+1} \ldots i_{-1} i_0 \cdot i_1 i_2 \ldots i_n \ldots}$ *and* $\mathcal{F}_{\ldots i_{-n} i_{-n+1} \ldots i_{-1} i_1 \cdot i_2 \ldots i_n \ldots}$ *for some sequence* $\ldots i_{-n} i_{-n+1} \ldots i_{-1} i_0 \cdot i_0 i_1 \ldots i_n \ldots$. *That is, the similarity map* φ *and the baker map* B *agree with each other.*

Each rectangle $\mathcal{F}_{i_{-k} i_{-n+1} \ldots i_{-1} i_0 \cdot i_1 \ldots i_n}$ *is made of labels for points of the intersection of the rectangles* $H_{i_{-k} i_{-n+1} \ldots i_{-1} i_0 \cdot}$ *and* $V_{\cdot i_1 \ldots i_n}$. *This immediately implies that the diameter condition is valid.*

The separation condition is also correct with $\varepsilon_0 = 1/4$ *and the second degree of separation. That is, hyperbolic domain structured chaos is approved for the baker transformation.*

We have suggested an abstraction for two-sided dynamical processes with discrete time. A method of chaos generation is provided as the dynamics of the abstract similarity map. Thus, all three types of chaos, Poincaré, Li–Yorke and Devaney, are proved in the domain. It is easy to see that homoclinic and heteroclinic points can be discussed in the framework of the current research. The approach also makes strong practical sense. For example, this has been realized in our papers [6, 7] for fractals and neural networks. Thus our comprehension of the dynamical complexities has been deepened. It is important that the chaotic behavior is extended for decreasing time, if one assumes the backward separation condition. This time we introduce the concept of abstract hyperbolic chaos, which has been verified for the Smale horseshoe map and the two-dimensional baker transformation. We are confident that the approach can shed more light on hyperbolic dynamics such as Anosov diffeomorphisms, Plykin attractors and others. A further benefit is multidimensional generalizations, which are possible on the basis of the proposals. Not only the deep past, but also the very far future have to be determined for the application of abstract similarity maps. That is, the hyperbolic dynamics considers processes with infinite chaotic history. However, the approach does not exclude the possibility of the description of random dynamics [4].

3.3 Modular chaos

In this section we apply the method of abstract similarity sets for a dynamics with orbits whirling among infinitely many modules. Thus the conservative conditions for the existence of chaos are weakened in the modeling of irregular dynamics. The research proves that Poincaré chaos is of exceptional use in the analysis of stochastic processes, and this will be verified in chapter 5. Examples illustrating the results are provided.

3.3.1 Introduction

The paradigm of chaos relies on the two distance antagonists, proximality and separation. They utilize metrical or topological closeness and divergence in the description considering either couples of orbits [26, 42] or a single motion [9]. In the current research we have renounced the idea that the transitivity and density of the periodic motions of a class of motions have to be described metrically, or even topologically. It is suggested, for the first time in the literature, that the closeness can be considered not in the metrical sense, but through *indexing*, although the separation still relies entirely on the distance. That is, the antagonists are still present, but not in the way of conventional theories. The process of indexing is adjacent to the similarity map. More precisely, the indexing is the instrument through which the mapping is to be chaotic. The trajectories move through different sets and even various metric spaces when the time increases and each of the sets admits its own chaotic structure. This is why we say that the *modular chaos* is the focus of this chapter or, more precisely, the *modular abstract similarity chaos*. Accordingly, if the chaos is indicated in models by labeling, we shall call it *modular domain structured chaos*.

The method of domain structured chaos was initiated in [4, 6, 7]. Results were obtained which rely on the dynamics in a single metrical or topological space, fractals and bounded random processes such as Bernoulli schemes [4, 7, 11]. This time the method has been enriched in two senses. First, we consider infinite multiples of chaotically structured modules. Second, the conservative chaos conditions are weakened by considering indexing for closeness. This helps us to determine the unbounded deterministic chaotic dynamics and find chaotic structures for continuous-time random processes, bounded as well as unbounded. We have agreed that chaos is present in discrete time stochastic dynamics if realizations coincide with orbits of the modular similarity map and explain its presence for continuous-time random processes (chapter 5). One must say that the concepts of Poincaré chaos and unpredictable sequences [9, 10] are central for the exploration.

We hope that the approach will be developed further by considering multidimensional indexing and infinitely dimensional dynamics. Additionally, the definitions can be considered for processes which are only partially stochastic, that is for hybrid dynamics, where intervals of deterministic and/or stochastic behavior are combined.

Let us shed more light on the probabilistic essence of the chaos and the random process relations. Considering the modular Poincaré chaos, one can see that there exists a modular unpredictable realization such that any finite realization is an arc of the unpredictable one. Consequently, we have to conclude that in any experiment we observe the modular unpredictable orbit as a certain event. This is the main argument as to why we compare the two types of dynamics in our research.

The research on similarity goes back to Leibniz, who introduced the notion of recursive self-similarity [64]. The idea of a self-similar set was first considered by Moran [44]. He gave a mathematical definition of a geometric construction as a collection of sets satisfying specific conditions. Further, definitions of self-similarity and related problems were discussed in [13, 27, 29, 35, 37–39, 46, 52, 58, 59]. Our research is based on a point-set structure in metric spaces and this is why we call it *abstract self-similarity* [7]. It does not rely on any special functions and the *similarity map* is naturally combined with the concept. The Bernoulli shift [61] is an example of the abstract similarity map. We assume that the next extension of the current results can be obtained on the basis of the papers [13, 27, 29, 35, 37, 39, 46, 58].

Potential applications of the modular chaos naturally relate to problems of control and synchronization [32]. The theory of stochastic differential equations [48] and Markov chains [30] can be enriched if modular chaos investigation results are involved in the study by utilizing methods of replication of chaos and unpredictable dynamics [5, 8, 9, 11].

It is of the strong interest to compare deterministic and random dynamics. More precisely, to find features of chaos in stochastic processes as well as to recognize probabilistic features in deterministic motions. If the first is considered, for example in papers [50, 51], in the paper [28] arguments of statistics are provided to show deterministic features in random processes. Apparently, our research is in the second stream, when the random processes can be assigned with properties of deterministic chaos. Our results rely not on the statistical parameters, but on the concept, introduced and developed in the last four years, of Poincaré chaos [9–11].

3.3.2 Modular similarity chaos

Let a countable collection of compact metric spaces (F_j, d_j), $j = 1, 2, \ldots$, be given with distances d_j. Assume that for every set F_j the following presentation exists:

$$\mathcal{F}^j = \left\{F^j_{i_1 i_2 \ldots i_n \ldots}: i_k = 1, 2, \ldots, m, \;\; k = 1, 2, \ldots\right\}, j = 1, 2, \ldots, \tag{3.19}$$

where m is a natural number not smaller than two and common for all these sets. This means that each element of the set F_j is labeled through at least one member of $\mathcal{F}^j$ and every element of set $\mathcal{F}^j$ presents a member from F_j. The uniqueness is not necessarily required for the relation. That is, if $\mathcal{F}^j_{i_1 i_2 \ldots i_n \ldots}$ and $\mathcal{F}^j_{j_1 j_2 \ldots jn, \ldots}$ present the same element f in F_j, it is not necessary that $i_n = j_n$ for all $n = 1, 2, \ldots$. The set $\mathcal{F}^j$ is said to be the *abstract similarity set* for F_j. Moreover, we determine functions $\delta_j(\mathcal{F}^j_{i_1 i_2 \ldots i_n \ldots}, \mathcal{F}^j_{j_1 j_2 \ldots jn, \ldots}) = d_j(f_1, f_2), j = 1, 2, \ldots,$ if $\mathcal{F}^j_{i_1 i_2 \ldots i_n \ldots}$ and $\mathcal{F}^j_{j_1 j_2 \ldots jn, \ldots}$ correspond to the elements f_1 and f_2 of the set F_j, respectively, such that $\delta_j(\mathcal{F}^j_{i_1 i_2 \ldots i_n \ldots}, \mathcal{F}^j_{j_1 j_2 \ldots jn, \ldots}) = 0$ for different presentations of the same point in F_j.

The following sets are needed:

$$\mathcal{F}^j_{i_1 i_2 \ldots i_n} = \bigcup_{j_k = 1, 2, \ldots, m} \mathcal{F}^j_{i_1 i_2 \ldots i_n j_1 j_2 \ldots}, \tag{3.20}$$

where indices $i_1, i_2, \ldots, i_n$, are fixed. It is clear that

$$\mathcal{F}^j \supseteq \mathcal{F}^j_{i_1} \supseteq \mathcal{F}^j_{i_1 i_2} \supseteq \ldots \supseteq \mathcal{F}^j_{i_1 i_2 \ldots i_n} \supseteq \mathcal{F}^j_{i_1 i_2 \ldots i_n i_{n+1}} \ldots, \;\; i_k = 1, 2, \ldots, m, \;\; k = 1, 2, \ldots.$$

That is, the sets form a nested sequence.

We will say that the *diameter condition* is valid for sets $\mathcal{F}^j_{i_1 i_2 \ldots i_n}, j = 1, \ldots,$ if

$$\sup_{i_1 i_2 \ldots i_n} \operatorname{diam} \mathcal{F}^j_{i_1 i_2 \ldots i_n} \to 0 \;\; \text{as} \;\; n \to \infty, \tag{3.21}$$

where $\operatorname{diam} A = \sup\{\delta_j(\mathbf{x}, \mathbf{y}): \mathbf{x}, \mathbf{y} \in A\}$, for a set A in $\mathcal{F}_j$.

Determine the function $\delta_j(A, B) = \inf\{\delta_j(\mathbf{x}, \mathbf{y}): \mathbf{x} \in A, \mathbf{y} \in B\}$, for two non-empty bounded sets A and B in $\mathcal{F}_j$. The set $\mathcal{F}^j, j = 1, 2, \ldots,$ satisfies the *separation condition* of degree n if there exist a positive number ε_0^j and a natural number $n = n(j)$ such that for arbitrary indices $i_1 i_2 \ldots i_n$ one can find indices $j_1 j_2 \ldots j_n$ such that

$$\delta_j\left(\mathcal{F}^j_{i_1 i_2 \ldots i_n}, \mathcal{F}^j_{j_1 j_2 \ldots j_n}\right) \geqslant \varepsilon_0^j. \tag{3.22}$$

If the diameter and separation conditions are valid, then $\mathcal{F}^j$ is said to be an *abstract chaotic set* for $F_j, j = 1, 2, \ldots,$ and an *abstract chaotic module* for F.

We shall say that the union, $\mathcal{F}$, of all modules $\mathcal{F}_j, j = 1, 2, \ldots,$ is a *abstract modular set* for the union F of all sets $F_j, j = 1, 2, \ldots,$ and it is called an *abstract modular chaotic set*, if the diameter condition is valid for all $j = 1, 2, \ldots$ and there exists a positive separation constant $\varepsilon_0 = \inf_{j=1,2,\ldots} \varepsilon_0^j$.

Let us introduce the maps $\varphi_j: \mathcal{F}^j \to \mathcal{F}^{j+1}, j = 1, 2, \ldots,$ such that

$$\varphi_j(\mathcal{F}^j_{i_1 i_2 \ldots i_n \ldots}) = \mathcal{F}^{j+1}_{i_2 i_3 \ldots i_n \ldots}. \tag{3.23}$$

In what follows we refer to the map φ on the set $\mathcal{F}$, which is specified by equations (5.22). That is, $\varphi(\mathcal{F}^j) = \varphi_j(\mathcal{F}^j), j = 1, 2, \ldots$. Moreover, accept that $\delta(\mathcal{F}^j_{i_1 i_2 \ldots i_n \ldots}, \mathcal{F}^j_{j_1 j_2 \ldots j_n \ldots}) = \delta_j(\mathcal{F}^j_{i_1 i_2 \ldots i_n \ldots}, \mathcal{F}^j_{j_1 j_2 \ldots j_n \ldots}), j = 1, 2, \ldots$.

Considering iterations of the map one can verify that

$$(\varphi_{j+n}(\dots(\varphi_j(\mathcal{F}^j_{i_1i_2\dots i_n}))\dots) = \varphi^n(\mathcal{F}^j_{i_1i_2\dots i_n})) = \mathcal{F}^{j+n+1} \tag{3.24}$$

for arbitrary natural number n and $i_k = 1, 2, \dots, m,\ k = 1, 2, \dots$. The relations (4.4) and (3.16) give us a reason to call φ the *affine similarity map*, since the shifts for the lower and upper indexes are in common. One can also call it the *modular similarity map*. Moreover, the number n is called the *order of similarity*. It is clear that the map cannot be considered as a modification of the Bernoulli map, since it is not a shift map. There is no direction in the phase space and shifts occur in the indices, but not in the elements of the dynamics. If $\mathcal{F}^j$ does not depend on j, the modular similarity map φ is known as the *similarity map* [4, 6].

We shall say that an element $\mathcal{F}^j_{i_1i_2\dots i_n\dots}$ is in the (k, m)-neighborhood of a point $\mathcal{F}^p_{s_1s_2\dots s_n\dots}$, $j < p$ if $\mathcal{F}^{j+k}_{i_{k+1}i_{k+2}\dots i_{k+m}} = F^p_{s_1s_2\dots s_m}$. In other words, $j + k = p$, $i_{k+1} = s_1, \dots, i_{k+m} = s_m$.

Since the concepts are based on the Bernoulli shift [61], known for the symbolic dynamics, it is easy to see that the following chaotic properties are valid.

A point $\mathcal{F}^j_{i_1i_2i_3\dots}$, $j = 1, 2, \dots$, is periodic with period p if its lower index consists of endless repetitions of a block of p terms. The map φ is said to be *modular-periodic*, since $\varphi^{kp}(\mathcal{F}^j_{i_1i_2\dots i_n\dots}) = \mathcal{F}^{j+kp}_{i_1i_2\dots i_n\dots}$ for any natural k, if the point $\mathcal{F}^j_{i_1i_2\dots i_n\dots}$ is p-periodic.

Similarly, the map is *modular-dense* or *modular-transitive*, since there exists the lower index $i_1i_2i_3\dots$, such that for any sequence $j_1j_2j_3\dots$ and natural j and l one can find positive integers m and $k, j < k$, such that $\varphi^m(\mathcal{F}^j_{i_1i_2\dots i_n\dots}) = \mathcal{F}^k_{j_1j_2\dots j_n\dots}$ and $i_s = j_s$, $s = 1, \dots, l$. In other words, $\mathcal{F}^j_{i_1i_2\dots i_n\dots}$ is in the (m, l)-neighborhood of $\mathcal{F}^k_{j_1j_2\dots j_n\dots}$.

Finally, one can say that the periodic points are *modular-dense* in $\mathcal{F}$, since for any point $\mathcal{F}^k_{j_1j_2\dots j_n\dots}$ of the set and natural numbers l and $m \leqslant k$ one can find a periodic point $\mathcal{F}^m_{i_1i_2\dots i_n\dots}$ such that it is in the $(k - m, l)$-neighborhood of $\mathcal{F}^k_{j_1j_2\dots j_n\dots}$.

It is clear that the couples $(\mathcal{F}^j, \delta_j)$, $j = 1, 2, \dots$, are not, in general, metric spaces. Nevertheless, the map φ is convenient to extend all attributes of Poincaré, Li–Yorke and Devaney chaos [11, 26, 42] for the dynamics, since the spaces (F_j, d_j) are metric. We shall say that the map φ is *modular chaotic in the Devaney sense* if it is dense with respect to modular-periodic trajectories, modular-transitive and *modular-sensitive.* The modular-sensitivity means that there exists a positive number ε_0 such that for each element $\mathcal{F}^j_{i_1i_2\dots i_n\dots}$ and arbitrary positive κ one can find an element $\mathcal{F}^j_{j_1j_2\dots j_n\dots}$ and a natural number k which satisfy $\delta(\varphi^k(\mathcal{F}^j_{i_1i_2\dots i_n\dots}), \varphi^k(\mathcal{F}^j_{j_1j_2\dots j_n\dots}) > \varepsilon_0$, despite $\delta(\mathcal{F}^j_{i_1i_2\dots i_n\dots}, \mathcal{F}^j_{j_1j_2\dots j_n\dots}) < \kappa$.

Let us determine the map Φ: $F_j \to F_{j+1}$, $j = 1, 2, \dots$, a geometrical presentation of the similarity map. Fix a number j and a point f from F_j. Then $\Phi(f)$ is the collection of all elements in F_{j+1}, the labels of which are values of the function φ at labels of f. It is clear that Φ is a set-valued mapping, but we shall call it a map for simplicity. In what follows, we shall describe the chaotic properties of the map Φ in terms of the dynamics of the map φ. The map Φ is *modular chaotic in the sense of Devaney*, if the corresponding map φ is modular chaotic in the same sense.

Theorem 3.3.1. *If $\mathcal{F}$ is an abstract modular chaotic set, then the dynamics of φ is modular chaotic in the sense of Devaney.*

Proof. Modular-transitivity and density for modular-periodicity are valid, due to the definitions. For modular-sensitivity fix a point $\mathcal{F}^j_{i_1 i_2\ldots} \in \mathcal{F}$ and an arbitrary positive number κ. Due to the diameter and separation conditions there exist an integer k and element $\mathcal{F}^j_{i_1\ldots i_k j_{k+1}\ldots}$ such that $0 < \delta(\mathcal{F}^j_{i_1\ldots i_k i_{k+1}\ldots}, \mathcal{F}^j_{i_1\ldots i_k j_{k+1}\ldots}) < \kappa$ and $\delta(\mathcal{F}^j_{i_{k+1}\ldots i_{k+n}}, \mathcal{F}^j_{j_{k+1}\ldots j_{k+n}}) > \varepsilon_0$. This proves the sensitivity. □

In [9, 10] Poisson stable motion is utilized to distinguish chaotic behavior from periodic motions in Devaney and Li–Yorke types of chaos. The dynamics is called Poincaré chaos. Let us provide the definition adapted for the dynamics of the current research. A point $\mathcal{F}^j_{i_1 i_2\ldots} \in \mathcal{F}$, $j = 1, 2, \ldots$, is said to be the *modular unpredictable point* for the map φ if there exist a positive number ε_0 and two unbounded sequences of natural numbers, κ_n and ζ_n, $n = 1, 2, \ldots$, such that for arbitrary natural number l one can find sufficiently large κ_n such that $\varphi^{\kappa_n}(\mathcal{F}^j_{i_1 i_2\ldots}) = \mathcal{F}^{j+\kappa_n}_{j_1 j_2\ldots}$, where $j_s = i_s$, $s = 1, \ldots, l$ and $j_{\kappa_n+\zeta_n} \neq i_{\kappa_n+\zeta_n}$.

Following the definition of Poincaré chaos [9] we shall say that the dynamics of the map φ is *modular chaotic in the sense of Poincaré* if there is a modular unpredictable point for each $j = 1, 2, \ldots$. Moreover, we call the unpredictable point itself an unpredictable orbit of the dynamics. It is natural to accept that the map Φ is *modular chaotic in the sense of Poincaré* if the corresponding map φ is modular chaotic in the sense of Poincaré.

Theorem 3.3.2. *If $\mathcal{F}$ is an abstract modular chaotic set, then the map φ possesses modular chaos in the sense of Poincaré.*

The proof of the last theorem is based on the verification of theorem 3.1.3 applied to the modular similarity map.

In addition to modular Devaney and Poincaré chaos, the chaos of Li–Yorke type [42] can be defined. As well as the modular-periodic points of the dynamics, assume that there exists an uncountable *scrambled set* set of non-periodic points in $\mathcal{F}^j$, $j = 1, 2, \ldots$, such that

$$(\alpha)\ \limsup_{k\to\infty} \delta(\varphi^k(\mathcal{F}^j_{i_1 i_2\ldots}), \varphi^k(\mathcal{F}^j_{i_j j_2\ldots})) > 0$$

and

$$(\beta)\ \liminf_{k\to\infty} \delta(\varphi^k(\mathcal{F}^j_{i_1 i_2\ldots}), \varphi^k(\mathcal{F}^j_{i_j j_2\ldots})) = 0$$

for each couple in the set, if $\delta(\mathcal{F}^j_{i_1 i_2\ldots}, \mathcal{F}^j_{i_j j_2\ldots}) \neq 0$. Finally, assume that for each point $\mathcal{F}^j_{i_1 i_2\ldots}$ of the set and a periodic point $\mathcal{F}^j_{i_j j_2\ldots}$ it is true that $\limsup_{k\to\infty} \delta(\varphi^k(\mathcal{F}^j_{i_1 i_2\ldots}),$

$\varphi^k(\mathcal{F}^j_{i_j j_2 \ldots})) > 0$. The conditions ($\alpha$) and ($\beta$) are called *modular-proximality* and *modular-frequent separation* properties for *the modular Li–Yorke chaos*, respectively.

We will say that for the sets $\mathcal{F}^j_{i_1 i_2 \ldots i_n}$, $j = 1, \ldots$, the *strong diameter condition* is valid, if

$$\sup_j \max_{i_1 i_2 \ldots i_n} \operatorname{diam} \mathcal{F}^j_{i_1 i_2 \ldots i_n} \to 0 \text{ as } n \to \infty. \tag{3.25}$$

We shall say that the union, $\mathcal{F}$, of all modules $\mathcal{F}_j$, $j = 1, 2, \ldots$, is a *strong abstract modular chaotic structure* for the union F of all sets F_j, $j = 1, 2, \ldots$, if the strong diameter condition is valid for all $j = 1, 2, \ldots$, and there exists a positive separation constant $\varepsilon_0 = \inf_{j=1,2,\ldots} \varepsilon_0^j$. We shall call the map Φ *modular chaotic in the sense of Li–Yorke*, if the corresponding map φ is modular chaotic in the sense of Li–Yorke.

Theorem 3.3.3. *If $\mathcal{F}$ is a strong abstract modular chaotic set, then the map φ is modular chaotic in the Li–Yorke sense.*

The proof of the assertion is similar to that of theorem 6.35 in [24] for the shift map defined in the space of symbolic sequences.

One can easily see that, in the case when (F_j, d_j) does not depend on j, we obtain domain structured chaos, which was introduced in [4]. That is, modular chaos can be considered as a generalization of domain structured chaos.

The analysis of the ingredients for the three types of chaos implies that the most weak and, consequently, most flexible properties are required for Poincaré modular chaos, since the distance evaluations are not required. Nevertheless, we are confident that this must not be considered as a deficiency. The description can still provide circumstances sufficient for irregularity, which can be useful for future applications. Thus, the following theorems can be of great interest.

Theorem 3.3.4. *If $\mathcal{F}$ is an abstract modular chaotic set, then the dynamics of Φ is modular chaotic in the sense of Devaney.*

Theorem 3.3.5. *If $\mathcal{F}$ is an abstract modular chaotic set, then the map Φ possesses modular chaos in the sense of Poincaré.*

Theorem 3.3.6. *If $\mathcal{F}$ is a strong abstract modular chaotic set, then the map Φ is modular chaotic in the Li–Yorke sense.*

3.3.3 Examples

This time we provide deterministic examples. Modular chaos for random processes is recognized in chapter 5.

Example 3.3.1. *Take into account metric spaces (F^j, d_j), $j = 1, 2, \ldots$, of sequences $f = (i_1^j, i_2^j, \ldots)$, the coordinates of which are elements of the finite sets $(a_1^j, a_2^j, \ldots, a_m^j)$, m is a natural number, and metrics*

$$d_j(f_1, f_2) = \sum_{l=1}^{\infty} \frac{\rho_j(i_l^j, k_l^j)}{2^{l-1}},$$

where $f_1 = (i_1^j, i_2^j, \ldots)$ *and* $f_2 = (k_1^j, k_2^j, \ldots)$ *are members of* F^j, $j = 1, 2, \ldots$, *and* ρ_j *are distances for elements of the alphabets. Considering modules* $\mathcal{F}^j$, $j = 1, 2, \ldots$, *as sets of labels* $\mathcal{F}^j_{i_1 i_2 \ldots} = (i_1^j, i_2^j, \ldots)$, *and the abstract similarity map we deal with modular chaos.*

Example 3.3.2. *It is known that there exists the Cantor set,* $\mathcal{C}$, *in the section* $[0, 1]$, *such that to each point of the set the label* $\tau_{i_1 i_2 \ldots}$, $i_k = 0, 1$, $k = 1, 2, \ldots$, *is assigned* [26, 61]. *Consider the structure* $\mathcal{T}$, *which consists of the labels and introduce the distance*

$$d(\tau_{i_1 i_2 \ldots}, \tau_{j_1 j_2 \ldots}) = |c_1 - c_2|,$$

where $|\cdot|$ *is the absolute value and* $\tau_{i_1 i_2 \ldots}$ *with* $\tau_{j_1 j_2 \ldots}$ *are labels for points* c_1 *and* c_2 *in the set. Consider shifts* $\mathcal{C}^j$, $j = 1, 2, \ldots$, *of the Cantor set* $\mathcal{C}$ *from the last example along the real axis by natural numbers. That is,* $\mathcal{C}^j = \{c + (j - 1) | c \in \mathcal{C}\}$, $j = 1, 2, \ldots$. *Accordingly, we label points of the Cantor sets as* $\tau^j_{i_1 i_2 \ldots}$, $i_k = 0, 1, k, j = 1, 2, \ldots$. *Determine the distance*

$$d_j(c_1, c_2) = d_j(\tau^j_{i_1 i_2 \ldots}, \tau^j_{j_1 j_2 \ldots}) = \sum_{k=1}^{\infty} \frac{|i_k - j_k|}{2^{k-1}}, j = 1, 2, \ldots ,$$

if $\tau^j_{i_1 i_2 \ldots}$ *and* $\tau^j_{j_1 j_2 \ldots}$ *are labels of the points* c_1 *and* c_2 *of the set* $\mathcal{C}^j$, $j = 1, 2, \ldots$. *By considering the similarity map*

$$\varphi(\tau^j_{i_1 i_2 \ldots}) = \tau^{j+1}_{i_2 i_3 \ldots}$$

we arrange the modular chaos on the union of the Cantor sets $\mathcal{C}^j$, *corresponding to modules* $\mathcal{T}^j$, $j = 1, 2, \ldots$. *To find a concrete meaning for the last construction, let us consider the non-autonomous scalar equation* $x_{i+1} = (i + 1) + m(x_i - i)(1 - x_i + i)$, $m > 4$. *Fix* $\mathcal{C}$ *as the invariant set for the map* $mx(1 - x)$ *in the unit section. It is clear that the union of sets* $\mathcal{C}^j$, $j = 1, 2, \ldots$, *is the domain for the modular chaos of the equations.*

3.4 Abstract fractals

We develop a new definition of fractals which can be considered as an abstraction of the fractals determined through self-similarity of label-sets. This time the idea of the domain structuring in a metric space is combined with the measuring of sets. The definition is formulated through imposing conditions which are governed by the relation between the subsets of a metric space to build a porous abstract self-similar structure. Examples are provided to confirm that the definition is satisfied by a large class of fractals determined by abstract self-similarity. The iterated function system is utilized to discover the structure of abstract fractals. The new concepts create new frontiers for fractals and the investigation of chaos.

3.4.1 Introduction

Fractals are a class of complex geometric shapes with certain properties. One of the main features of the objects is self-similarity, which can be defined as the property whereby parts hold similarity to the whole at any level of magnification [2]. The roots of the idea of self-similarity date back to the seventeenth century when Leibniz introduced the notions of recursive self-similarity [64]. The first mathematical definition of a self-similar shape was introduced in 1872 by Karl Weierstrass during his study of functions that were continuous but not differentiable. The most famous examples of fractals that display exact self-similarity are the Cantor set, Koch curve, and Sierpinski gasket and carpet, which were discovered by Georg Cantor in 1883, Helge von Koch in 1904 and Waclaw Sierpinski in 1916, respectively. Julia sets, discovered by Gaston Julia and Pierre Fatou in 1917–19, gained significance in being generated using the dynamics of iterative functions. In 1979 Mandelbrot visualized Julia sets, including a fractal called the Mandelbrot set. The fractional dimension is suggested by Mandelbrot to be a property of fractals when he defined a fractal as a set whose Hausdorff dimension is strictly larger than its topological dimension [43].

In this paper we introduce a new mathematical concept and call it the *abstract fractal*. This concept is an attempt to establish a pure foundation for fractals by abstracting the idea of self-similarity. We define the abstract fractal as a collection of points in a metric space. The points are represented through an iterative construction algorithm with specific conditions. The conditions are introduced to govern the relationship between the sets at each iteration. Our approach of construction is based on the concept of porosity rather than the roughness notion introduced by Mandelbrot. Porosity is an intrinsic property of materials and it is usually defined as the ratio of void volume to total volume [12]. The concept of porosity plays an important role in several fields of research such as geology, soil mechanics, material science, civil engineering, etc [12, 31]. Fractal geometry has been widely used to study the properties of porous materials. However, the concept of porosity was not utilized as a criterion for fractal structures and the relevant researchers have investigated the relationship between porosity and fractalness [22, 25, 34, 36, 63]. For example, several studies such as [54, 60, 62] determined the fractal dimension of some pore-structures using their properties. The simplicity and importance of the porosity concept insistently invites us to develop a new definition of fractals through porosity. In other words, the property should be involved in fractal theory as a feature to be equivalent to self-similarity and fractional dimension. This needs to specify the concept of porosity to surfaces and lines. In the present chapter we do not pay attention to the equivalence between the definition of fractals in terms of porosity and those through self-similarity and dimension, instead we introduce an abstract definition which, we hope, will be useful in application domains.

3.4.2 The definition

In this part we shall consider the metric measure space defined by the triple (X, d, μ), where (X, d) is a compact metric space, d is a metric on X and μ is a measure on X.

To construct an abstract fractal, let us consider the initial set $F \subset X$ and fix two natural numbers m and M such that $1 < m < M$. We assume that there exist M nonempty disjoint subsets, F_i, $i = 1, 2, \ldots M$, such that $F = \cup_{i=1}^{M} F_i$. For each $i = 1, 2, \ldots m$, again, there exist M nonempty disjoint subsets F_{ij}, $j = 1, 2, \ldots M$ such that $F_i = \cup_{j=1}^{M} F_{ij}$. Generally, for each $i_1, i_2, \ldots, i_n$, $i_k = 1, 2, \ldots m$, there exist M nonempty disjoint sets $F_{i_1 i_2 \ldots i_n j}$, $j = 1, 2, \ldots M$, such that $F_{i_1 i_2 \ldots i_n} = \cup_{j=1}^{M} F_{i_1 i_2 \ldots i_n j}$ for each natural number n. The following conditions are required. There exist two positive numbers, r and R, such that for each natural number n we have

$$r \leqslant \frac{\sum_{j=1}^{m} \mu\left(F_{i_1 i_2 \ldots i_{n-1} j}\right)}{\sum_{j=m+1}^{M} \mu\left(F_{i_1 i_2 \ldots i_{n-1} j}\right)} \leqslant R, \tag{3.26}$$

where $i_k = 1, 2, \ldots, m$, $k = 1, 2, \ldots n - 1$. We call the relation (3.26) the *ratio condition*. The numbers r and R in equation (3.26) are characteristics for porosity. Another condition is the *adjacent condition* and it is formulated as follows. For each $i_1 i_2 \ldots i_n$, $i_k = 1, 2, \ldots, m$ there exists j, $j = m + 1, m + 2, \ldots, M$, such that

$$d(F_{i_1 i_2 \ldots i_n}, F_{i_1 i_2 \ldots i_{n-1} j}) = 0. \tag{3.27}$$

We call $F_{i_1 i_2 \ldots i_{n-1} i_n}$ a complement set of order n if $i_k = 1, 2, \ldots, m$, $k = 1, 2, \ldots n - 1$ and $i_n = m + 1, m + 2, \ldots, M$.

An accumulation point of any couple of complement sets does not belong to any of them. We dub this stipulation the *accumulation condition*.

Let us define the diameter of a bounded subset A in X by $\mathrm{diam}(A) = \sup\{d(\mathbf{x}, \mathbf{y}): \mathbf{x}, \mathbf{y} \in A\}$. Considering the above construction, we assume that the *diameter condition* holds for the sets $F_{i_1 i_2 \ldots i_n}$, i.e.

$$\sup_{i_k = 1, 2, \ldots M} \mathrm{diam}\, F_{i_1 i_2 \ldots i_n} \to 0 \ \text{as} \ n \to \infty. \tag{3.28}$$

Fix an infinite sequence $i_1 i_2 \ldots i_n \ldots$. The diameter conditions as well as the compactness of X imply that there exists a sequence (p_n), such that $p_0 \in F$, $p_1 \in F_{i_1}$, $p_2 \in F_{i_1 i_2}$, $\ldots$, $p_n \in F_{i_1 i_2 \ldots i_n}$, $n = 1, 2, \ldots$, which converges to a point in X. The points are denoted by $F_{i_1 i_2 \ldots i_n \ldots}$.

We define the *abstract fractal* $\mathcal{F}$ as the collection of the points $F_{i_1 i_2 \ldots i_n \ldots}$ such that $i_k = 1, 2, \ldots m$, that is

$$\mathcal{F} = \{F_{i_1 i_2 \ldots i_n \ldots} \mid i_k = 1, 2, \ldots m\}, \tag{3.29}$$

provided that the above four conditions hold. The subsets of $\mathcal{F}$ can be represented by

$$\mathcal{F}_{i_1i_2\ldots i_n} = \left\{F_{i_1i_2\ldots i_ni_{n+1}i_{n+2}\ldots} \mid i_k = 1, 2, \ldots, m\right\}, \tag{3.30}$$

where $i_1i_2\ldots i_n$ are fixed numbers. We call such subsets *subfractals* of order n. It is clear that an abstract fractal is an abstract self-similar set and there is abstract similarity chaos, if the separation condition is fulfilled. This problem is under discussion in chapter 4.

3.4.3 Abstract fractals in geometry

In this subsection we find the pattern of an abstract fractal in some geometrical well-known fractals: the Sierpinski carpet, Pascal triangles and Koch curve.

3.4.3.1 The Sierpinski carpet

To construct an abstract fractal corresponding to the Sierpinski carpet, let us consider a square as an initial set F. First, we divide F into nine ($M = 9$) equal squares and denote them by F_i, $i_1 = 1, 2, \ldots 9$ (see figure 3.7(a)). In the second step, each square F_i, $i = 1, 2, \ldots 8$ is again divided into nine equal squares denoted as F_{ij}, $j = 1, 2, \ldots 9$. Figure 3.7(b) illustrates the sub-squares of F_1. We continue in this way such that at the nth step, each set $F_{i_1i_2\ldots i_{n-1}}$, $i_k = 1, 2, \ldots 8$, is divided into nine subsets $F_{i_1i_2\ldots i_{n-1}j}$, $j = 1, 2, \ldots 9$. For the Sierpinski carpet the number m is 8, and the measure ratio (3.26) can be evaluated as follows. If we consider the first order sets F_{i_1}, $i_1 = 1, 2, \ldots 9$, then

$$\frac{\sum_{j=1}^{8} \mu(F_j)}{\mu(F_9)} = 8.$$

Thus the ratio condition holds. From the construction we can see that each $F_{i_1i_2\ldots i_n}$, $i_k - 1, 2, \ldots, 8$ a has common boundary with $F_{i_1i_2\ldots i_{n-1}j}$, $j = 9$. Therefore the adjacent condition holds. Since the construction consists of division into smaller parts, the diameter condition is also valid. Moreover, it is clear that the accumulation condition holds as well.

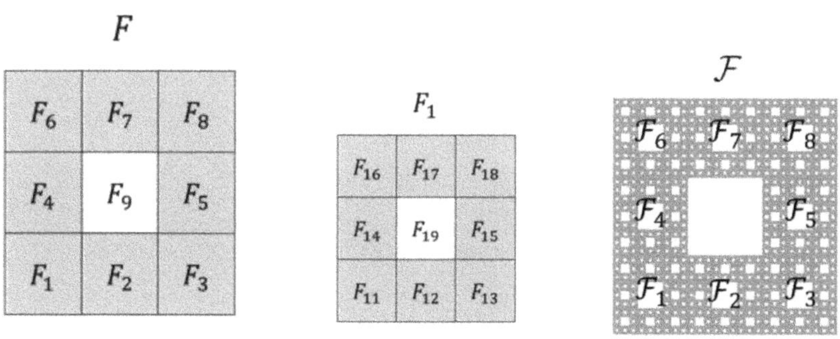

(a) (b) (c)

Figure 3.7. Sierpinski carpet.

As a result the points of the desired abstract fractal $\mathcal{F}$ can be represented as $F_{i_1 i_2 \ldots i_n \ldots}$ and the abstract Sierpinski carpet is defined by

$$\mathcal{F} = \{F_{i_1 i_2 \ldots i_n \ldots} | i_k = 1, 2, \ldots 5\}.$$

Figure 3.7(c) shows the set $\mathcal{F}$ and illustrates its first order subfractals.

3.4.3.2 The Pascal triangle

The Pascal triangle is a mathematical structure consisting of a triangular array of numbers. Triangular fractals can be obtained if these numbers are plotted using specific moduli. The Sierpinski gasket, for example, is the Pascal's triangle modulo 2. Let us build an abstract fractal on the basis of a fractal associated with Pascal's triangle modulo 3. Consider an equilateral triangle as an initial set F. In the first step we divide F into nine smaller equilateral triangles and denote them as F_i, $i = 1, 2, \ldots, 9$, as shown in figure 3.8(a). Next, each triangle F_i, $i = 1, 2, \ldots, 6$ is again divided into nine equilateral triangles named F_{ij}, $j = 1, 2, \ldots, 9$. Figure 3.8(a) illustrates the second step for the set F_1. Similarly, the subsequent steps are performed such that at the nth step, each set $F_{i_1 i_2 \ldots i_{n-1}}$, $i_k = 1, 2, \ldots 6$, is divided into nine subsets $F_{i_1 i_2 \ldots i_{n-1} j}$, $j = 1, 2, \ldots 9$. In this case we have $m = 6$ and $M = 9$. Therefore

$$\frac{\sum_{j=1}^{6} \mu(F_j)}{\sum_{j=7}^{9} \mu(F_j)} = 2$$

and the ratio condition holds. One can also verify that the adjacent, the accumulation and the diameter conditions are also valid. Based on this, the points of the fractal can be defined by $F_{i_1 i_2 \ldots i_n \ldots}$ and thus the abstract Pascal triangle is defined by

$$\mathcal{F} = \{F_{i_1 i_2 \ldots i_n \ldots} | i_k = 1, 2, \ldots 6\}$$

and the nth order subfractals can be written as

$$\mathcal{F}_{i_1 i_2 \ldots i_n} = \{F_{i_1 i_2 \ldots i_n i_{n+1} \ldots} \mid i_k = 1, 2, \ldots, 6\},$$

where $i_1 i_2 \ldots i_n$ are fixed numbers.

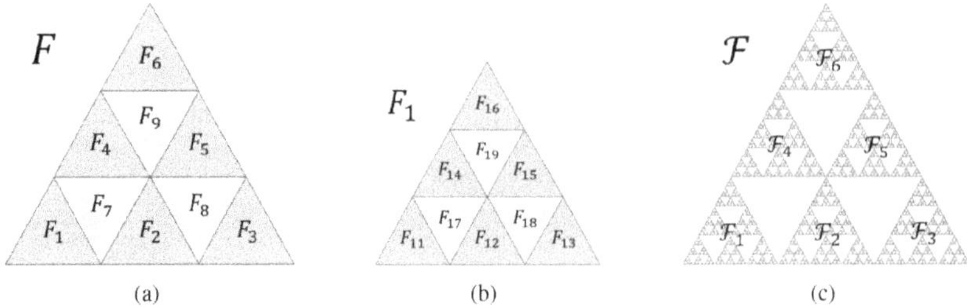

Figure 3.8. Pascal triangle modulo 3.

3.4.3.3 The Koch curve

In this subsection we shall show how to build an abstract fractal $\mathcal{F}$ conformable to the Koch curve. For this purpose we consider the following construction of the Koch curve. Start with an isosceles triangle F with base angles of 30°. The first step of the construction consists in dividing F into three equal-area triangles F_1, F_2 and F_3 (see figure 3.9(b)). The triangles F_1 and F_2 are isosceles with base angles of 30°, whereas the central triangle F_3 is an equilateral one. In the second step each F_i, $i = 1, 2$, is similarly divided into three triangles, two isosceles, F_{i1} and F_{i2}, and one equilateral, F_{i3}. Figure 3.9(c) illustrates this step. In each subsequent step the same procedure is repeated for each isosceles triangle resulting from the preceding step. That is, in the nth step each $F_{i_1 i_2 \ldots i_{n-1}}$, $i_k = 1, 2$, is divided into three parts, two isosceles triangles $F_{i_1 i_2 \ldots i_{n-1} j}$, $j = 1, 2$, with base angles of 30°, and one equilateral triangle $F_{i_1 i_2 \ldots i_{n-1} 3}$. In this construction we have $m = 2$ and $M = 3$, thus the measure ratio is

$$\frac{\mu(F_1) + \mu(F_2)}{\mu(F_3)} = 2$$

and the ratio condition holds. From the construction it is clear that the adjacent, the accumulation and the diameter conditions are also valid. Based on this the points in $\mathcal{F}$ can be represented as $F_{i_1 i_2 \ldots i_n \ldots}$ and thus the abstract Koch curve is defined by

$$\mathcal{F} = \{F_{i_1 i_2 \ldots i_n \ldots} \mid i_k = 1, 2\}.$$

The nth order subfractals of $\mathcal{F}$ are represented by

$$\mathcal{F}_{i_1 i_2 \ldots i_n} = \left\{F_{i_1 i_2 \ldots i_n i_{n+1} i_{n+2} \ldots} \mid i_k = 1, 2\right\}, \tag{3.31}$$

where $i_1 i_2 \ldots i_n$ are fixed numbers. Figure 3.10 illustrates examples of the first, second, third and fourth subfractals of the abstract Koch curve.

3.4.4 Fractals by an iterated function system

An iterated function system (IFS) is a powerful tool for the construction of fractal sets. It is defined by a family of contraction mappings w_n, $n = 1, 2, \ldots N$ on a

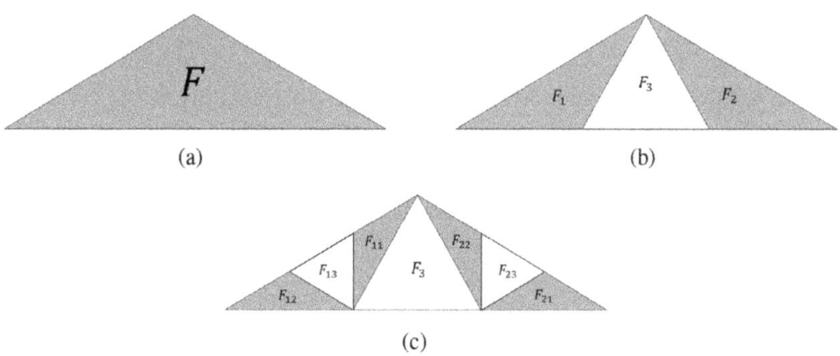

Figure 3.9. Abstract Koch curve construction.

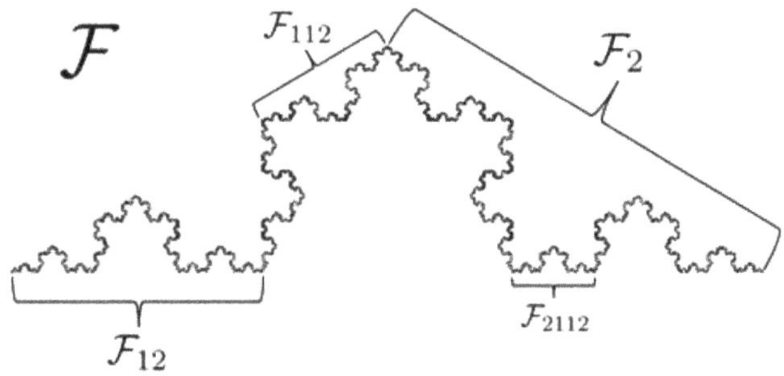

Figure 3.10. Subfractals of the abstract Koch curve.

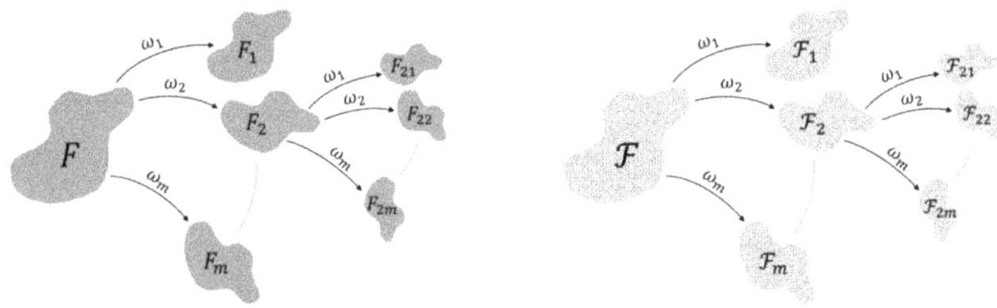

(a) (b)

Figure 3.11. Abstract fractals and the iterated function system.

complete metric space (X, d) [15, 37]. The procedure starts with choosing an initial set $A_0 \in \mathcal{B}(X)$, where $\mathcal{B}(X)$ is the space of the nonempty compact subsets of X, then iteratively applying the map $W = \{w_n, n = 1, 2, \ldots N, \}$ such that $A_{k+1} = W(A_k) = \bigcup_{n=1}^{N} A_k^n$, where $A_k^n = w_n(A_k)$. The fixed point of this map, $A = W(A) = \lim_{k\to\infty} W^k(A_0) \in \mathcal{B}(X)$, is called the attractor of the IFS which represents the intended fractal.

The idea of the structure of the abstract fractal can be realized using IFS. The fractal constructed by an IFS is an invariant set. Therefore the subsets at each step of construction can be determined using the maps w_n, $n = 1, 2, \ldots m$, as illustrated in figure 3.11(a). Similarly, the maps transform each subfractal into subfractals of the subsequent order. Figure 3.11(b) demonstrates the action of w_ns on the abstract fractal. The difference between this case and the above IFS fractal construction is that the sets $\mathcal{F}_{i_1 i_2 \ldots i_n}$ are fractals in themselves, whereas the sets A_k^n are not.

Utilizing this idea, moreover, each subfractal can be expressed in terms of the iterated images of the whole fractal $\mathcal{F}$, that is

$$\mathcal{F}_i = w_i(\mathcal{F}), \quad \mathcal{F}_{ij} = w_j(w_i(\mathcal{F})), \quad \mathcal{F}_{ijk} = w_k(w_j(w_i(\mathcal{F}))), \ \ldots ,$$

thus in general we have

$$\mathcal{F}_{i_1 i_2 \ldots i_n} = w_{i_n}(w_{i_{n-1}}(\ldots(w_{i_1}(\mathcal{F}))\ldots),$$

from which we can define a point belonging to the fractal as the limit of the iterated images of $\mathcal{F}$,

$$\mathcal{F}_{i_1 i_2 \ldots} = \lim_{n \to \infty} w_{i_n}(w_{i_{n-1}}(\ldots(w_{i_1}(\mathcal{F})))\ldots).$$

The existence of a separation constant ε_0 can be expressed in terms of w_n such that the condition is satisfied if

$$\min_{n} \inf_{i_n, j_n = 1, 2, \ldots m} d\big(w_{i_n}(\ldots(w_{i_1}(\mathcal{F})), w_{j_n}(\ldots(w_{j_1}(\mathcal{F}))\big) \geqslant \varepsilon_0.$$

In addition to the construction of fractals, the IFS is used to prove chaos for the so-called totally disconnected IFS corresponding to certain classes of self-similar fractals such as the Cantor set [14]. The proof consists of the construction of a dynamical system $\{A; S\}$, where $S: A \to A$ is the shift transformation defined by $S(a) = w_n^{-1}(a)$ for $a \in W_n(A)$. The system is called the shift dynamical system associated with the IFS and it is shown to be topologically conjugate to the shift map on the N-symbols code space. We see that this approach follows the usual construction of chaos which begins with defining a map with certain properties where the conjugacy to a well-known chaotic map is the major key to discover the chaotic nature of fractals. Again we emphasize that this approach is only applicable for the totally disconnected fractals, namely the well-known Cantor sets. Differently, our approach is characterized by the similarity map $\varphi: \mathcal{F} \to \mathcal{F}$, which, with regard to the IFS approach, can be seen as an abstraction of the geometric essence of the transformation S. Using the idea of indexing the domain elements allows us to define the abstract map φ. This has shortened the method of proving chaos by eliminating the need for topological conjugacy. Moreover, it becomes possible to investigate the chaotic nature in several classes of fractals such as the Sierpinski fractals and the Koch curve.

3.5 Notes

In our research the abstraction of the self-similarity concept is now accomplished. In fact, it is the main concept which is the basis of the current research. It takes us on another new level of the Pythagoras (c. 570 BC–495 BC) paradigm of natural numbers as fundamentals of everything. If one looks for simplicity, the ten basic digits are the fundamentals. Our research is still seeking a combination of simplicity and universality, and perhaps the abstract similarity is the concept which issues the tools which are needed and are designed for the study of modern dynamics and sophisticated motions. The first members of the toolkit are the abstract similarity map, acting in abstract similarity sets, as well as abstract hyperbolic similarity sets, modular similarity sets and the notion of abstract fractals. Next, to adapt the set of tools for research needs, we formulated conditions such as the separation and diameter conditions. In summary, the abstract similarity dynamics concept has been shaped. The separation condition guarantees sensitivity of the dynamics and the diameter condition is an auxiliary one to make the required closeness possible. It is needless to say that the conditions are the sufficient and necessary for chaos. Specifically, unpredictability, the Poisson stability, proximality and frequency separation are present in domains, if they are assigned with abstract similarity sets through labeling. We have revisited traditional models such as baker transformations, two- and several-fold, one- and two-dimensional, the symbolic dynamics, the logistic map and Smale horseshoes to demonstrate how the domain structured chaos method is effective and simple in chaotic dynamics investigations. More progressive examples are given in the following chapters. We strongly believe that one can shed more light on the dynamics, which satisfies the conditions of the Ornstein isomorphism theorem [49] and prove that period three implies domain structured chaos [42] by application of the results of abstract similarity dynamics.

The fractal concept is axiomatically linked with the notion of self-similarity. This is why it is considered to be one of the two acceptable definitions of fractals. That is, a fractal can be defined as a set that displays self-similarity at all scales. Mandelbrot defined a fractal as a set whose Hausdorff dimension is strictly larger than its topological dimension. In the current research, we introduce the concept of an abstract fractal which can be considered as another criterion of fractalness. Indeed, the idea of the abstract fractal centers around the self-similarity property and many self-similar fractals such as the Cantor sets and the Sierpinski fractals are shown to be fractals in the sense of the abstract fractal. These fractals also satisfy the Mandelbrot definition. Moreover, in our previous paper [7] we have also shown that the set of symbolic strings satisfies the definition of abstract self-similarity. Because of these facts we believe that the notion of the abstract fractal deserves to be the third definition of fractals and we hope it will be accepted by the mathematical community. Considering the abstract fractal as a new definition of fractals may open new opportunities for more theoretical investigations in this field as well as new possible applications in science and engineering. For example, we may start with the equivalency between these definitions. It is known that the fractals that display exact self-similarity at all scales satisfy the Mandelbrot definition of a fractal. The proposed definition satisfies the self-similarity since it is the main pivot of the concept of the abstract fractal. However, the interesting question is: does the abstract fractal agree with the Mandelbrot definition? The notion of Hausdorff dimension for the abstract fractal is not yet developed enough to provide an answer to the question. However, the determination of the fractal dimension can possibly be performed based on two important properties. The first is the self-similarity of the abstract fractal which may provide a self-similar dimension that can be assumed to be equivalent to the Hausdorff dimension. The second is the accumulation condition combined perhaps with the diameter condition. These properties are essential for describing the geometry of fractals, therefore the fractal dimension can be characterized in terms of them. This is why, the definition in our paper can give opportunities to compare abstract fractals with fractals defined through dimension. Furthermore, the suggested fractal definition can be elaborated through chaotic dynamics development, topological spaces, physics, chemistry and the development of neural network theories.

References

[1] Addabbo T, Fort A, Rocchi S and Vignoli V 2011 Digitized chaos for pseudo-random number generation in cryptography *Chaos-based Cryptography: Theory, Algorithms and Applications* ed L Kocarev and S Lian (Berlin: Springer) pp 67–97
[2] Addison P S 1997 *Fractals and Chaos: an Illustrated Course* (Bristol, UK: Institute of Physics Publishing)
[3] Adler R, Konheim A and McAndrew M 1965 Topological entropy *Trans. Am. Math. Soc.* **114** 309–19
[4] Akhmet M 2019 Domain-structured chaos in discrete random processes arXiv:1912.10478 [math.DS]
[5] Akhmet M 2020 Domain structured chaos in Markov chains arXiv: 2010.14625v1 [math.DS]
[6] Akhmet M and Alejaily E M 2019 Domain-structured chaos in a Hopfield neural network *Int. J. Bifurcat. Chaos* **29** 1950205
[7] Akhmet M and Alejaily E M 2021 Abstract similarity, chaos and fractals *Discrete Contin. Dyn. Syst.* B **22** 2479–97
[8] Akhmet M and Fen M O 2016 *Replication of Chaos in Neural Networks, Economics and Physics* (Berlin: Springer)
[9] Akhmet M and Fen M O 2016 Poincaré chaos and unpredictable functions *Commun. Nonlinear Sci. Numer. Simul.* **48** 85–94
[10] Akhmet M and Fen M O 2016 Unpredictable points and chaos *Commun. Nonlinear Sci. Numer. Simul.* **40** 1–5
[11] Akhmet M, Fen M O and Alejaily E M 2020 *Dynamics with Chaos and Fractals* (Berlin: Springer)
[12] Anovitz L M and Cole D R 2015 Characterization and analysis of porosity and pore structures *Rev. Mineral. Geochem.* **80** 61–164
[13] Bandt C and Graf S 1992 Self-similar sets 7. A characterization of self-similar fractals with positive Hausdorff measure *Proc. Am. Math. Soc.* **114** 995–1001
[14] Barnsley M 1988 *Fractals Everywhere* (New York: Academic)
[15] Barnsley M F and Demko S 1985 Iterated function systems and the global construction of fractals *Proc. R. Soc.* A **399** 243–75
[16] Bowen R 1970 Markov partitions and minimal sets for axiom A diffeomorphisms *Am. J. Math.* **92** 907–18
[17] Bowen R 1970 Markov partitions for axiom A diffeomorphisms *Am. J. Math.* **92** 725–47
[18] Bowen R 1973 Symbolic dynamics for hyperbolic flows *Am. J. Math.* **95** 429–60
[19] Bowen R 1975 A horseshoe with positive measure *Invent. Math.* **29** 203–4
[20] Bowen R and Ruelle D 1975 The ergodic theory of axiom A flows *Invent. Math.* **29** 181–202
[21] Bunimovich L and Ya S 1980 Markov partitions for dispersed billiards *Commun. Math. Phys.* **78** 247–80
[22] Cai J C, Yu B M, Zuo M Q and Mei M F 2010 Fractal analysis of surface roughness of particles in porous media *Chin. Phys. Lett.* **27** 024705
[23] Cartwright M and Littlewood J 1945 On nonlinear differential equations of the second order I: the equation $y'' - k(1 - y^2)y' + y = bk\cos(\lambda t + a)$, k large *J. London Math. Soc.* **20** 180–9
[24] Chen G and Huang Y 2011 *Chaotic Maps: Dynamics, Fractals and Rapid Fluctuations, Synthesis Lectures on Mathematics and Statistics* (San Rafael, CA: Morgan and Claypool)

[25] Davis H T 1989 On the fractal character of the porosity of natural sandstone *Europhys. Lett.* **8** 629–32
[26] Devaney R L 1989 *An Introduction to Chaotic Dynamical Systems* (Canada; USA: Addison-Wesley)
[27] Edgar G A 1990 *Measure, Topology, and Fractal Geometry* (New York: Springer)
[28] Eisencraft M, Monteiro L H A and Soriano D C 2017 White Gaussian chaos *IEEE Commun. Lett.* **21** 1719–22
[29] Falconer K J 1995 Sub-self-similar sets *Trans. Am. Math. Soc.* **347** 3121–9
[30] Gagniuc P A 2017 Markov chains: from theory to implementation and experimentation *Markov Chains: From Theory to Implementation and Experimentation* (Hoboken, NJ: Wiley)
[31] Ganji D D and Kachapi S H H 2015 *Application of Nonlinear Systems in Nanomechanics and Nanofluids: Analytical Methods and Applications* (New York: Elsevier)
[32] Gonzáles-Miranda J M 2004 *Synchronization and Control of Chaos* (London: Imperial College Press)
[33] Guckenheimer J and Holmes P J 1997 *Nonlinear Oscillations, Dynamical Systems and Bifurcations of Vector Fields* (New York: Springer)
[34] Guyon E, Mitescu C D, Hulin J P and Roux S 1989 Fractals and percolation in porous media and flows? *Physica* D **38** 172–8
[35] Hata M 1985 On the structure of self-similar sets *Japan J. Appl. Math.* **2** 381–414
[36] Huang H 1993 Porosity-size relationship of drilling mud flocs: fractal structure *Clays Clay Miner.* **41** 373–9
[37] Hutchinson J 1981 Fractals and self-similarity *Indiana Univ. Math. J.* **30** 713–47
[38] Jorgensen P E T 2006 *Analysis and Probability: Wavelets, Signals, Fractals. Graduate Texts in Mathematics* vol 234 (New York: Springer)
[39] Lau K S, Ngai S M and Rao H 2001 Iterated function systems with overlaps and the self-similar measures *J. London Math. Soc.* **63** 99–116
[40] Levinson N 1949 A second order differential equation with singular solutions *Ann. Math.* **50** 127–53
[41] Li R and Zhou X 2013 A note on chaos in product maps *Turk. J. Math.* **37** 665–75
[42] Li T Y and Yorke J A 1975 Period three implies chaos *Am. Math. Monthly* **82** 985–92
[43] Mandelbrot B B 1983 *The Fractal Geometry of Nature* (New York: Freeman)
[44] Moran P A P and Yorke J A 1946 Additive functions of intervals and Hausdorff measure *Proc. Camb. Philos. Soc.* **42** 15–23
[45] Newhouse S 1980 *Lectures on Dynamical Systems, Progress in Mathematics* (New York: Birkhauser)
[46] Ngai S M and Wang Y 2001 Hausdorff dimension of overlapping self-similar sets *J. London Math. Soc.* **63** 665–72
[47] Nicolis G and Prigogine I 1989 *Exploring Complexity: An Introduction* (New York: Freeman)
[48] Oksendal B K 2003 *Stochastic Differential Equations: An Introduction with Applications* (Berlin: Springer)
[49] Ornstein D 1970 Bernoulli shifts with the same entropy are isomorphic *Adv. Math.* **4** 337–52
[50] Ornstein D 1995 In what sense can a deterministic system be random? *Chaos Solitons Fractals* **5** 139–41
[51] Ornstein D 2004 Kolmogorov, random processes, and Newtonian dynamics *Russ. Math. Surv.* **59** 313–7

[52] Pesin Y and Weiss H 1996 On the dimension of deterministic and random Cantor-like sets, symbolic dynamics, and the Eckmann–Ruelle conjecture *Commun. Math. Phys.* **182** 105–53
[53] Poincaré H 1957 *New Methods of Celestial Mechanics* vol 1–3 (New York: Dover)
[54] Puzenko A, Kozlovich N, Gutina A and Feldman Y 1999 Determination of pore fractal dimensions and porosity of silica glasses from the dielectric response at percolation *Phys. Rev.* B **60** 14348–59
[55] Ruelle D 1978 An inequality of the entropy of differentiable maps *Bull. Braz. Math. Soc.* **9** 83–7
[56] Smale S 1966 Diffeomorphisms with many periodic points *Differential and Combinatorial Topology: A Symposium in Honor of Marston Morse* (Princeton, NJ: Princeton University Press) pp 63–70
[57] Smale S 1967 Differentiable dynamical systems *Bull. Am. Math. Soc.* **73** 747–817
[58] Spear D W 1992 Measure and self-similarity *Adv. Math.* **91** 143–57
[59] Stella S 1992 On Hausdorff dimension of recurrent net fractals *Proc. Am. Math. Soc.* **116** 389–400
[60] Tang H P, Wang J Z, Zhu J L, Ao Q B, Wang J Y, Yang B J and Li Y N 2012 Fractal dimension of pore-structure of porous metal materials made by stainless steel powder *Powder Technol.* **217** 383–7
[61] Wiggins S 1988 *Global Bifurcation and Chaos, Analytical Methods* (New York: Springer)
[62] Xia Y, Cai J, Wei W, Hu X, Wang X I N and Ge X 2018 A new method for calculating fractal dimensions of porous media based on pore size distribution *Fractals* **26** 1850006
[63] Yu B and Liu W 2004 Fractal analysis of permeabilities for porous media *AIChE J.* **50** 46–57
[64] Zmeskal O, Dzik P and Vesely M 2013 Entropy of fractal systems *Comput. Math. Appl.* **66** 135–46

Chapter 4

Domain structured dynamics: applications

Sections of this chapter have been reprinted from [3], copyright (2021), with permission from AIMS.

The projection on a Bernoulli shift along an infinite sequence of symbols is one of the most powerful tools in the modern study of non-linearity. However, there is a strong deficiency—one has to establish the projection as an auxiliary mapping and that is evidently not an easy task. In chapter 3 we provided a deeply developed method of structuring state spaces such that the rule of motion between the labeled elements reflects the required dynamical features directly. In this chapter we present to our reader pioneering results of the application of the method. The theory of the domain structured dynamics emerged through applications. The algorithm appeared primarily in our research of dynamics in fractals. Then it was found that the approach of arranging chaos is suitable for non-fractal figures. Surprisingly, the traditional chaotic models such as the logistic map and baker transformation can be clarified much more easily and elegantly than previously if abstract similarity dynamics is utilized. This is why we emphasize that our method of exploration of dynamical features as well as the arrangement of sophisticated motions is convenient for first-time readers. The advantages for entirely new investigations are provided by the confirmation of chaos in the majority of fractals determined by self-similarity, the construction of unpredictable functions, finding new laws for discrete random processes, as well as the observation of chaotic features in discrete and continuous time stochastic motions. We shed light on the application potential considering the following aspects: advantages of the method for enlargement of the span of chaotic domains; constructive algorithms of labeling; and fractal growth as the interaction of progressive and conservative features in development. This will all be covered in the current and subsequent chapters.

4.1 Theoretical preliminaries

We start with basic definitions and theorems. Assume that (G, d) is a compact metric space with the distance d. Assume that there exists the following presentation for the set G:

$$F = \{F_{i_1 i_2 \ldots i_n \ldots} : i_k = 1, 2, \ldots , m, \ k = 1, 2, \ldots\}, \tag{4.1}$$

where m is a natural number. In this chapter the presentation means that each element of the set G is labeled through a single element of the set F. We keep the distance d for the set F considering $d(F_{i_1 i_2 \ldots i_n \ldots}, F_{j_1 j_2 \ldots jn, \ldots}) = d(g_1, g_2)$, if $F_{i_1 i_2 \ldots i_n \ldots}$ and $F_{j_1 j_2 \ldots jn, \ldots}$ are labels of elements g_1 and g_2 of the set G. In what follows we call the set F a *label set* for the set G. We shall denote by $\mathcal{L}$ the labeling of the set G such that $\mathcal{L}$: $G \to F$. It is clear that the map is a bijection. More precisely, it is an isometry. The following sets, which are called rectangles, are required:

$$F_{i_1 i_2 \ldots i_n} = \bigcup_{j_k = 1, 2, \ldots, m} F_{i_1 i_2 \ldots i_n j_1 j_2 \ldots}, \tag{4.2}$$

where indices $i_1, i_2, \ldots , i_n$, are fixed.

It is clear that

$$F \supseteq F_{i_1} \supseteq F_{i_1 i_2} \supseteq \ldots \supseteq F_{i_1 i_2 \ldots i_n} \supseteq F_{i_1 i_2 \ldots i_n i_{n+1}} \ldots, \ i_k = 1, 2, \ldots , m, \ k = 1, 2, \ldots.$$

That is, the rectangles form a nested sequence. Let us introduce the map φ: $F \to F$ such that

$$\varphi(F_{i_1 i_2 \ldots i_n \ldots}) = F_{i_2 i_3 \ldots i_n \ldots}. \tag{4.3}$$

Considering iterations of the map, one can verify that

$$\varphi^n(F_{i_1 i_2 \ldots i_n}) = F, \tag{4.4}$$

for arbitrary natural number n and $i_k = 1, 2, \ldots , m, \ k = 1, 2, \ldots , n$. The relations (4.3) and (4.4) give us a reason to call φ the *abstract similarity map* or just the *similarity map*. We shall call the triple (F, φ, d) *abstract similarity dynamics.* Introduce the function ψ: $G \to G$ such that $\psi(g) = \mathcal{L}^{-1}(\varphi(\mathcal{L}(g)))$. The triple (G, ψ, d) is said to be the *domain structured dynamics.*

We will say that for rectangles $F_{i_1 i_2 \ldots i_n}$ the *diameter condition* is valid if

$$\sup_{i_k = 1, 2, \ldots, m} \operatorname{diam} F_{i_1 i_2 \ldots i_n} \to 0 \ \text{as} \ n \to \infty, \tag{4.5}$$

where $\operatorname{diam} A = \sup\{d(\mathbf{x}, \mathbf{y}): \mathbf{x}, \mathbf{y} \in A\}$ for a set A in F.

Consider the function $d(A, B) = \inf\{d(\mathbf{x}, \mathbf{y}): \mathbf{x} \in A, \mathbf{y} \in B\}$ of two nonempty sets A and B in F. The space (F, d) satisfies the *separation condition* of degree n if there exist a positive number ε_0 and a natural number n such that for arbitrary indices $i_1 i_2 \ldots i_n$ one can find indices $j_1 j_2 \ldots j_n$ such that

$$d\left(F_{i_1 i_2 \ldots i_n}, F_{j_1 j_2 \ldots j_n}\right) \geqslant \varepsilon_0. \tag{4.6}$$

If the diameter and separation conditions are valid, then F is said to be a *chaotic structure* for G. A point $F_{i_1 i_2 i_3 \ldots}$ from F is periodic with period n if its index consists of endless repetitions of a block of n terms. The following theorems are valid.

Theorem 4.1.1. [3] *If F is a chaotic structure, then the similarity map possesses Poincaré chaos.*

Theorem 4.1.2. [3] *If F is a chaotic structure, then the similarity map is chaotic in the sense of Devaney.*

Theorem 4.1.3. [3] *If F is a chaotic structure, then the similarity map possesses Li–Yorke chaos.*

Then, the isometry of $\mathcal{L}$ it implies that the following theorems are valid.

Theorem 4.1.4. [3] *If the separation and diameter conditions are valid, then the dynamics (G, ψ, d) possesses Poincaré chaos.*

Theorem 4.1.5. [3] *If the separation and diameter conditions are valid, then the dynamics (G, ψ, d) is chaotic in the sense of Devaney.*

Theorem 4.1.6. [3] *If the separation and diameter conditions are valid, then the dynamics (G, ψ, d) possesses Li–Yorke chaos.*

The chaos for the triple (G, ψ, d) is said to be *domain structured chaos*.

The proofs of the last six theorems can be found in chapter 3 and in [3]. Thus, if one says that domain structured chaos is present in the space (G, ψ, d) it means that the three types of chaos, Poincaré, Li–Yorke and Devaney, occur.

The method of domain structured dynamics is a very general tool of theoretical analysis and it is a new, higher level of abstraction of the research. In this chapter we shall use it as *structural dynamization*, when a metric space is not given a dynamics. By considering the domain structured dynamics (F, φ, d) chosen properly for a metric space, we are able to establish a desired behavior in the space, for example chaos, or more precisely domain structured chaos, the presence of which is guaranteed by the abstract similarity chaos. First, this has been shown for fractals. We believe that the result is of significant interest, since it is a demonstration that fractals can be chaotic domains. Moreover, it is shown in our research that chaos can be arranged in finite-dimensional cubes and consequently for any set which is topologically equivalent to the cubes. Another surprise is that chaos can be formulated for finite sets and even two-point sets. This creates nice opportunities to look for the ingredients of chaos in stochastic processes. Generally speaking, the approach is a fundamental contribution to recognize chaos as the most generic type of motion in the world. This is very important for seeking chaotic dynamics in neural networks, since it is needed for the effectiveness of brain activity and artificial intelligence. Another significant field of applications is computer science, when one transforms a static data structure into a dynamic one [41]. Finally, we recommend paying attention to the conservative–progressive method for fractal growth, presented in section 4.4. It is distinct from the discussion of fractal growth in the physical sense [48] as we consider mathematical aspects of the problem, possibly for the first time in the literature.

Fatou–Julia iterations are also a target for the application of domain structured dynamics. The center of the approach is the brilliant idea of Julia and Fatou to collect bounded orbits of recurrent dynamics, indicating them approximately, and thus fractals, which are necessarily self-similar sets. We have applied the iterations as an instrument of labeling such that the domain structured dynamics admits a new direction. Thus the famous types of fractals have been assigned domain structured dynamics and chaos. If they are confirmed to have separation and diameter conditions, then domain structured chaos is seen. Otherwise, we still are looking for the conditions. Thus for Julia and Mandelbrot sets the problem has to be solved and chaos verified. Accordingly, we believe that abstract similarity sets can be determined for the fractal sets such that the self-similarity will be proved in the abstract way.

4.2 Chaotic finite-dimensional cubes

In this section we show that a chaotic behavior can be found on cubes with arbitrary finite dimension. The method of domain structured chaos is applied. Thus we shall prove that a cube can be a quasi-minimal set. Previously this was known only for the unit section and for Devaney and Li–Yorke chaos.

4.2.1 Introduction

Chaos has become a very important concept that is deeply integrated in many, if not all, fields of science such as physics, biology, medicine and engineering, and human culture and activities [13, 46]. The chaotic behavior of some physical and biological properties was formerly attributed to random or stochastic processes or uncontrolled forces [29, 34]. The appearance of chaos in deterministic systems drew the border between deterministic chaos and stochastic noise. The idea is manifested in the chaotic behavior of simple dynamical systems. However, the randomness theory of Kolmogorov–Martin–Löfwhich can still provide a deeper understanding of the origins of deterministic chaos [13]. The fundamental theoretical framework of chaos was developed in the last quarter of the twentieth century. During that period, different types and definitions of chaos were formulated. In general, chaos can be defined as aperiodic long-term behavior in a deterministic system that exhibits sensitive dependence on the initial conditions [28]. Devaney [17] and Li–Yorke [32] chaos are the most frequently used types and are characterized by transitivity, sensitivity, frequent separation and proximality. Another common type occurs through a period-doubling cascade, which is a route to chaos through local bifurcations [19, 43, 44]. In the papers [5, 6] Poincaré chaos was introduced through the unpredictable point concepts and it was developed further to unpredictable functions and sequences.

Whoever studies in this field can discern from the literature that there is a scientific conception that chaos is everywhere. Realizing such an ideation needs our mathematical tools to be developed to conceptualize all manifestations of the phenomenon. Strictly speaking, we should develop simple chaotic mechanisms that have the ability to emulate complex behaviors. Investigating the fundamental aspects of high-dimensional chaotic states is necessary in this direction. Indeed, mathematical modeling of real-world problems shows that real life is very often a high-dimensional chaos and even chaotic activities in our everyday lives are difficult to describe via low-dimensional systems [25].

Recently, in papers [2, 3], we have developed a new method of chaos formation which depends instead on the approach of partitioning of the domain, rather than on a map. That is, the map is a natural consequence of the structure of the domain being chaotic. In the current study we extend the approach of considering the methodological problem for chaos as a *generic* dynamical phenomenon in the real world, science and industry. This problem has not been discussed in our previous research and the current study is complementary to those in [2, 3]. The phenomenon is understood in two aspects. The first is connected to the number of models which admit chaotic dynamics. This problem is difficult to discuss since the number of differential, discrete and other equations exhibiting chaos is still vanishingly small, if you compare them to the number of equations admitting regular behavior. The second aspect concerns the density of chaotic points in the domain or the state space of the dynamics. This question has not been considered in detail (as far as we know) in the literature. This is because the chaotic domains are usually considered as fractal objects, that is, their dimension is less than the dimension of the state space. Consequently, it is assumed without discussion that the points of chaos are sparse and the domain admits a small density only for one-dimensional unimodal dynamics, as is the case for the unit section where it was proven that all points of the set can be chaotic [17, 32]. As a first step towards the goal, we show in the present work how to establish multidimensional chaos for simple geometrical objects. We focus on the domain structure to construct an invariant set under a chaotic map. Our approach is based on a recursive division of the set into an infinite number of elements which satisfy specific conditions. By using infinite sequences to index the set points, a chaotic map can be defined on the set. Thus, what we are doing in this chapter is the realization of domain structured dynamics for simple geometrical objects. It is clear that the results are valid for all figures which are topologically equivalent to cubes.

4.2.2 Chaotic line segment

For the sake of comprehension, let us start with the line segment $G = (0, 1]$. Divide the set into four parts $F_1 = \left[0, \frac{1}{4}\right]$, $F_2 = \left(\frac{1}{4}, \frac{1}{2}\right]$, $F_3 = \left(\frac{1}{2}, \frac{3}{4}\right]$ and $F_4 = \left(\frac{3}{4}, 1\right]$ (see figure 4.1 (b)). Divide each part F_{i_1}, $i_1 = 1, 2, 3, 4$, again into four equal parts and denote them by $F_{i_1 i_2}$, $i_2 = 1, 2, 3, 4$. Continue with this procedure such that, at the kth step of the partition, each part $F_{i_1 i_2 \ldots i_{k-1}}$ is divided into four equal parts denoted as $F_{i_1 i_2 \ldots i_k}$, $i_p = 1, 2, 3, 4$, $p = 1, 2, \ldots, k$. Figure 4.1(c) illustrates the second step of the partition of the part F_1.

Now one can see that each point of the unit section is an intersection of nested lines F_{i_1}, $F_{i_1 i_2}$, … , and it is reasonable to label the point as $F_{i_1 i_2 \ldots i_k \ldots}$, such that

$$F = \left\{F_{i_1 i_2 \ldots i_k \ldots} \mid i_p = 1, 2, 3, 4, \; p = 1, 2, \ldots\right\}. \tag{4.7}$$

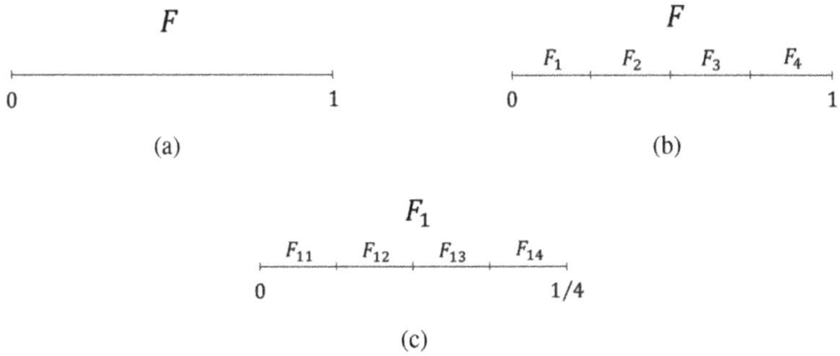

Figure 4.1. The first three labeling steps of the unit section.

That is, the F is a *label set* for the unit section in the sense of domain structured dynamics determined in the last chapter. It is sensible to consider the line with an Euclidean metric. Then the length of a section $F_{i_1i_2\ldots i_k}$, $k = 1, 2, \ldots$, is equal to $1/4k$ and the diameter condition is obviously valid. Similarly, we can see that the first degree separation condition is true with $\varepsilon_0 = 1/4$. Indeed, $d(F_1, F_3) = d(F_1, F_4) = d(F_2, F_4) = \frac{1}{4}$.

Let us now consider the similarity map φ: $F \to F$ defined by

$$\varphi(F_{i_1i_2\ldots i_k\ldots}) = F_{i_2i_3\ldots i_k\ldots}, \tag{4.8}$$

such that for a fixed sequence $i_1i_2\ldots i_k$, $\varphi(F_{i_1i_2\ldots i_k}) = F_{i_2i_3\ldots i_k}$ and $\varphi(F_{i_1}) = F$. We call each part $F_{i_2i_3\ldots i_k}$ a *rectangle of order k*.

Considering the results in section 4.1 one can agree that the similarity map is chaotic in the sense of Devaney, Li–Yorke and Poincaré. Thus we have shown that the line segment is a domain for chaos. This simple case is clearly pointed out in [17] for the Devaney chaos of the logistic map '$f(x) = 4x(1 - x)$' on the interval [0, 1].

To realize the similarity map φ geometrically one can consider the following four-fold scalar baker transformation ψ:

$$\psi(x) = \begin{cases} 4x & 0 < x \leqslant \dfrac{1}{4}, \\ 4\left(x - \dfrac{1}{4}\right) & \dfrac{1}{4} < x \leqslant \dfrac{1}{2}, \\ 4\left(x - \dfrac{1}{2}\right) & \dfrac{1}{2} < x \leqslant \dfrac{3}{4}, \\ 4\left(x - \dfrac{3}{4}\right) & \dfrac{3}{4} < x \leqslant 1. \end{cases}$$

We define the first order subsets as $F_1 = \left(0, \frac{1}{4}\right]$, $F_2 = \left(\frac{1}{4}, \frac{1}{2}\right]$, $F_3 = \left(\frac{1}{2}, \frac{3}{4}\right]$, $F_4 = \left(\frac{3}{4}, 1\right]$, and for each $i = 1, 2, 3, 4$ we have $G = \{\psi(x): x \in F_i\}$. For each

$j = 1, 2, 3, 4$ the second order subsets are defined by $F_{ij} = \{x \in F_i\colon \psi(x) = F_j\}$. Generally, a kth order subset can be defined by

$$F_{i_1 i_2 \dots i_k} = \{x \in F_{i_1 i_2 \dots i_{k-1}}\colon \psi(x) = F_{i_2 i_3 \dots i_k}\},$$

where $i_p = 1, 2, 3, 4,\ p = 1, 2, \dots, k$. We can easily see that the four-fold baker map agrees with the similarity map. This is why it is the easiest way to conclude that the baker transformation is domain structured chaotic and, consequently, there is Poincaré, Li–Yorke and Devaney chaos.

4.2.3 Chaotic square and cube

On the basis of the above construction, more general cases of chaotic domains can be investigated. Consider the unit square (cube) G, $(0, 1]^2$, $((0, 1]^3)$. Similarly to the line case, we divide G into 16 squares (64 cubes) and denote them by F_{i_1}, $i_1 = 1, 2, \dots, 16$ (F_{i_1}, $i_1 = 1, 2, \dots, 64$). Illustrations of the first step of construction for a square and cube are seen in figure 4.2. Again we divide each square (cube) F_{i_1} into 16 squares (64 cubes) and denote them by $F_{i_1 i_2}$, $i_2 = 1, 2, \dots, 16$ ($F_{i_1 i_2}$, $i_2 = 1, 2, \dots, 64$). We continue this procedure such that at the kth step of partition each part $F_{i_1 i_2 \dots i_{k-1}}$ is divided into 16 squares (64 cubes) denoted as $F_{i_1 i_2 \dots i_k}$, $i_p = 1, 2, \dots, 16$, $p = 1, 2, \dots, k$ ($F_{i_1 i_2 \dots i_k}$, $i_p = 1, 2, \dots, 64$, $p = 1, 2, \dots, k$). Likewise, the label set F can be defined by

$$F = \left\{F_{i_1 i_2 \dots i_k \dots} \mid i_p = 1, 2, \dots, m,\ p = 1, 2, \dots\right\},$$

where m is 16 for the square and 64 for the cube. Analogously, in these cases the similarity map φ is defined and it is chaotic in the sense of Devaney, Li–Yorke and Poincaré.

4.2.4 Multidimensional cubes

For a general case, consider the unit n-dimensional cube $G = (0, 1]^n$. The first step consists of dividing F into 4^n equal parts (n-dimensional sub-cubes) denoted as

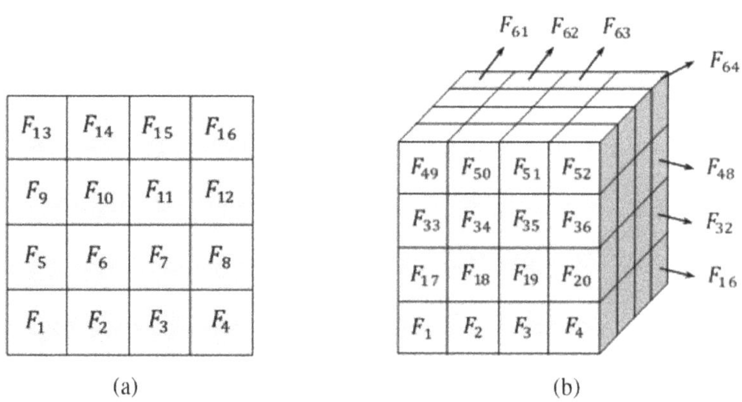

Figure 4.2. The partition procedure of a square and cube.

F_{i_1}, $i_1 = 1, 2, \ldots, 4^n$. In the second step each part F_{i_1} is again divided into 4^n equal parts denoted as $F_{i_1 i_2}$, $i_2 = 1, 2, \ldots, 4^n$. Continue in this procedure such that at the kth step of partition each part $F_{i_1 i_2 \ldots i_{k-1}}$ is divided into 4^n equal sub-cubes denoted as $F_{i_1 i_2 \ldots i_k}$, $i_p = 1, 2, \ldots, 4^n$, $p = 1, 2, \ldots, k$. At an infinite iteration of this process, all points of the cube G can be represented as the collection of labels $F_{i_1 i_2 \ldots i_k \ldots}$, i.e. the label set is the following collection,

$$F = \left\{F_{i_1 i_2 \ldots i_k \ldots} \mid i_p = 1, 2, \ldots, 4^n, \ p = 1, 2, \ldots\right\}. \tag{4.9}$$

Let us now verify the diagonal and separation properties stated in section 4.1. At any step k the diagonal of the sub-cube $F_{i_1 i_2 \ldots i_k}$ is given by $\frac{\sqrt{n}}{4^k}$ and, since n is finite, the diagonal property holds.

For the separation property we will show that it is valid with a separation constant $\varepsilon_0 = \frac{\sqrt{n}}{4}$. Consider an edge of the cube as a projection of the cube on the dimension. The partition allows us to find several sub-cubes such that the distance between them and a sub-cube A is not less than $\frac{1}{4}$ in the projection. Consequently, there exists a sub-cube, say B, which is distanced from A in all projections not less than $\frac{1}{4}$. This is why the distance between A and B is greater than or equal to $\frac{\sqrt{n}}{4}$. Here, we point out that the separation constant has a direct relationship with the dimension of the cube. This simply means that the high dimensionality makes the chaos stronger and this agrees with the concept that higher state space dimensions allow for more complex attitudes including chaotic behavior [23].

In figure 4.3 we provide simulation results that illustrate the chaotic behavior of trajectories initiated at points on the unit square and cube under the similarity map φ.

The proposed approach for generating chaotic dynamics on infinite sets sheds light on the rule of the domain of chaos. In the classical theory the requirements and properties of chaos are usually described through the map and the chaotic behavior is then reflected in the image which is characterized by some quantitative measures such as fractional dimension, similarity and scaling. An example of such a case is the Lorenz attractor whose capacity dimension is 2.06 ± 0.01 [33]. The same could be

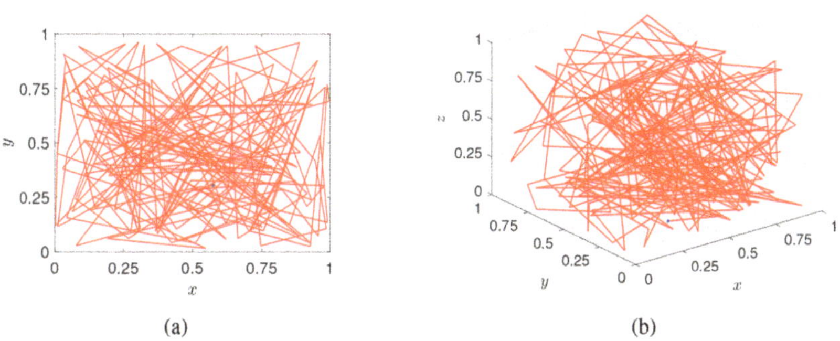

Figure 4.3. The chaotic behavior of trajectories: (a) initiated at the point ≈(0.574 633 735 9, 0.302 773 856 5) on the unit square and (b) initiated at the point ≈(0.376 643 795 1, 0.264 931 823 0, 0.027 599 813 5) on the unit cube.

said for an invariant subset of the domain such as a Cantor set for the logistic map which has the Hausdorff dimension 0.630 93 [36]. In the current research we devote more attention to the domain and less attention to properties of the map. In particular, we impose specific conditions on the domain, namely the diagonal and separation properties. On the other hand, the continuity and injectivity requirements for the motion are ignored since a chaotic map need not be continuous or injective [1, 18, 27, 31, 42]. We regard these properties of a chaotic map as important characteristics only from the analytical side, that is to say, they are very useful for handling the map to prove the presence of chaos [16, 51], however, it is not rigorously correct to consider them as intrinsic properties for chaos. Disregarding such properties helps us to organize a given infinite set as an invariant domain of the map which leads to a chaotic regime with the domain, the map and the image having equal dimensions. This is different from the classical situation in continuous chaotic dynamics where the dimension of the map is usually larger than that for the domain and image as in the case of the above two examples, the Lorenz system and the logistic map.

The results presented in this section reveal that the points of an n-dimensional cube are chaotic under the defined map and it is clear that the approach can be extended for each infinite set which admits topological properties common to the cube. This reinforces the concept that the chaotic points are dense in the state space of dynamics of many real-world phenomena. Moreover, the results would be crucial for numerous relevant analyses and applications. For example, a good application can be performed for Brownian motion [14, 49] where the random movement of particles may be considered in light of the present results.

4.3 Similarity through Fatou–Julia iterations

The brilliant idea of the famous scientists Fatou and Julia [26] is the approximation of infinite time dynamics in a bounded area with long orbits. The method does not have advantages for regular motions, but is very useful for complex behavior and modern computational powers allow this approach to flourish for applications. Currently, the simple and powerful algorithm is a unique method for the construction of sophisticated geometries. In the present section we use the Fatou–Julia iterations for the domain structuring.

4.3.1 Systems of logistic maps

Let $X = \mathbb{R}^n$ and $F_0 = F = [0, 1]^n$ is the n-dimensional unit cube. Consider the n-dimensional logistic map $\psi = (\psi_1, \psi_2, \dots \psi_n)$: $\mathbb{R}^n \to \mathbb{R}^n$ defined by

$$\begin{aligned} x_{k+1}^1 &= \psi_1(x_k^1) = r_1 x_k^1(1 - x_k^1), \\ x_{k+1}^2 &= \psi_2(x_k^2) = r_2 x_k^2(1 - x_k^2), \\ &\vdots \\ x_{k+1}^n &= \psi_n(x_k^n) = r_n x_k^n(1 - x_k^n), \end{aligned} \tag{4.10}$$

where $r_i > 4$, $i = 1, 2, \dots, n$, are parameters.

In [3] we adapt the idea of Fatou–Julia iteration (FJI), sometimes called the 'escape time algorithm' (ETA), and develop a scheme for constructing the Sierpinski carpet. The scheme is based on the iterations of the modified planar tent map. The Sierpinski fractals are typically generated using an iterated function system (IFS) [10, 11], which is defined as a collection of affine transformations. The technique of the FJI is based on detecting the points of a fractal set through the limits of their iterations under a specific map. Here we shall extend the technique to include any possible criterion for grouping points in a given domain. That is, domain structuring will be realized. Proceeding upon the properties of the logistic map, the ETA is applied to the map (4.10). We iterate the points in F under the map (4.10) such that in each iteration we keep only the points whose images do not escape the domain F_0. The resulting points from the first iteration belong to the subsets F_i, $i = 1, 2, \ldots, 2^n$. In the second iteration the non-escaped points belong to 2^{2n} subsets and each subset is indexed as F_{ij}, $j = 1, 2, \ldots, 2^n$, such that $F_{ij} \subset F_i$ and $\psi(F_{ij}) = F_j$. Similarly, a subset resulting at the kth iteration is indexed as $F_{i_1 i_2 \ldots i_k}$ such that $F_{i_1 i_2 \ldots i_k} \subset F_{i_1 i_2 \ldots i_{k-1}}$ and $\psi(F_{i_1 i_2 \ldots i_k}) = F_{i_2 i_3 \ldots i_k}$.

Based on the algorithm, it is clear that the separation condition holds. Thus we describe the points $F_{i_1 i_2 i_3 \ldots} = \lim_{k \to \infty} F_{i_1 i_2 \ldots i_k}$ and then the self-similar set F corresponding to the above algorithm is defined as the collection of the points $F_{i_1 i_2 i_3 \ldots}$. It is clear that the map ψ and similarity map ϕ coincide on F.

For r_i, $i = 1, 2, \ldots, n$, larger than 4, the separation condition of the first degree and diameter condition are guaranteed to be valid for the set F and, therefore, theorems 4.1.2, 4.1.1 and 4.1.3 imply that the similarity map ϕ and thus ψ are chaotic in the sense of Poincaré, Li–Yorke and Devaney.

For the numerical simulation, let us consider the two-dimensional system

$$\begin{aligned} x_{n+1} &= \psi_1(x_n) = r_1 x_n(1 - x_n), \\ y_{n+1} &= \psi_2(y_n) = r_2 y_n(1 - y_n), \end{aligned} \tag{4.11}$$

with $r_1 = 4.2$ and $r_2 = 4.3$. We fix $G = [0, 1] \times [0, 1]$ and apply ETA for equation (4.11). The first iteration will generate the sets F_i, $i = 1, 2, 3, 4$. In the second iteration we obtain the sets F_{ij}, $j = 1, 2, 3, 4$, and so on. Figure 4.4 shows the subsets constructed in the first three iterations. The chaotic set of the system (4.11) corresponding to the label set is that resulting from an infinite iteration of this procedure. The set is a type of Cantor dust which is the Cartesian product of two Cantor sets [30].

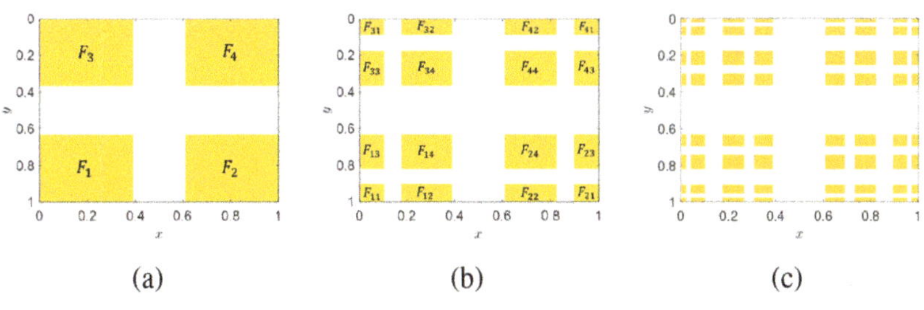

Figure 4.4. The construction of an abstract self-similar set using the map (4.11). Reprinted from [8]. Copyright (2020) with permission of Springer.

4.3.2 Perturbed systems

Let F_0 denote the initial set $[0, 1] \times [0, 1]$ and consider the two-dimensional perturbed logistic map $\psi = (\psi_1, \psi_2)$: $\mathbb{R}^2 \to \mathbb{R}^2$ defined by

$$\begin{aligned} x_{n+1} &= \psi_1(x_n, y_n; \mu_1) = r_1 x_n(1 - x_n) + \mu_1 y_n, \\ y_{n+1} &= \psi_2(x_n, y_n; \mu_2) = r_1 y_n(1 - y_n) + \mu_2 x_n, \end{aligned} \tag{4.12}$$

where r_1, r_2, μ_1 and μ_2 are parameters. The last term on the left-hand side of both equations in system (4.12) can be considered as a perturbation of a unimodal map. It is known that if a unimodal map is regular [24, 52], then it is structurally stable and, therefore, any small perturbation does not affect the topological properties of the map [9]. Such an inference can be extended for high-dimensional unimodal maps. Numerical simulations can provide an adequate verification of the unimodal properties of the perturbed map.

Similarly to subsection 4.3.1, we apply ETA to the map (4.12). The points that do not escape F_0 in the first iteration belong to the subsets F_i, $i = 1, 2, 3, 4$. In the second iteration the resulting points belong to the subsets indexed by F_{ij}, $j = 1, 2, 3, 4$, such that $F_{ij} \subset F_i$ and $\psi(F_{ij}) = F_j$. Similarly, a subset resulting at the nth iteration is indexed as $F_{i_1 i_2 \ldots i_n}$ such that $F_{i_1 i_2 \ldots i_n} \subset F_{i_1 i_2 \ldots i_{n-1}}$ and $\psi(F_{i_1 i_2 \ldots i_n}) = F_{i_2 i_3 \ldots i_n}$. Figure 4.5 shows the subsets constructed in the first three iterations with the parameters $r_1 = 4.2$, $r_2 = 4.5$, $\mu_1 = 0.03$ and $\mu_2 = -0.05$.

Depending on the choice of the coefficients r_i and relying on the smallness of the coefficients μ_i, we have that the diameter and separation conditions for abstract similarity and chaos are fulfilled. Therefore, we could say that the similarity map ϕ agrees with equation (4.12) and is chaotic on the self-similar-set F as a union of labels $F_{i_2 i_3 \ldots}$. Figure 4.6 depicts two trajectories which start at the points (0.044 608 921 784 357, 0.287 657 531 506 301) and (0.910 182 036 407 281, 0.329 865 973 194 639). The irregular behavior of the trajectories reveals the presence of chaos in equation (4.12).

Consider the space $X = \mathbb{R}^n$ and let F be a compact set in X such that it contains an open neighborhood of the n-dimensional unit cube and the set F_0 to be sufficiently near to the cube. Consider the n-dimensional perturbed logistic map $\psi = (\psi_1, \psi_2, \ldots \psi_n)$: $\mathbb{R}^n \to \mathbb{R}^n$ which is defined by

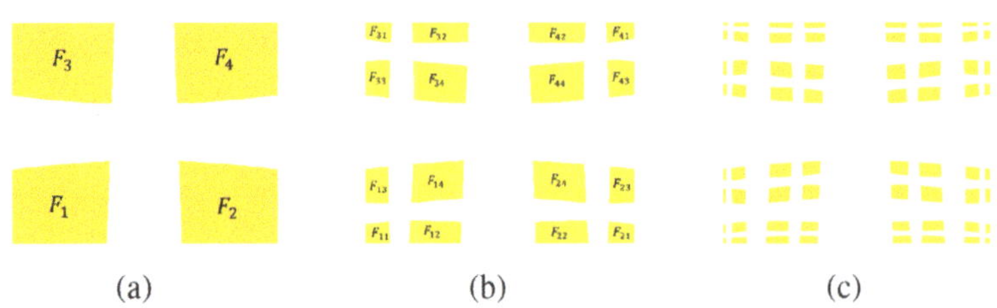

Figure 4.5. The first three iterations of the self-similar set construction using the map (4.12). Reprinted from [8]. Copyright (2020) with permission of Springer.

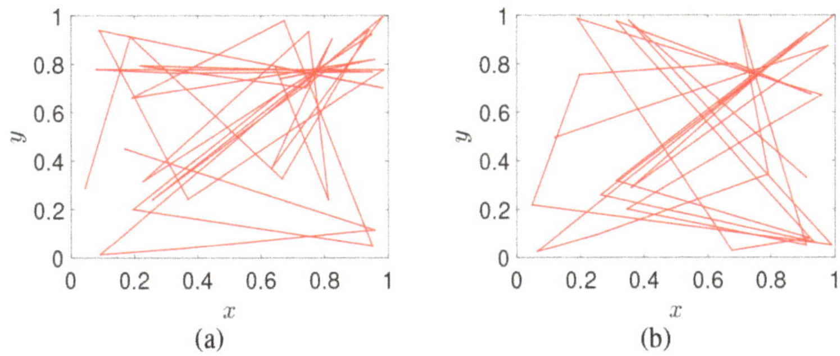

Figure 4.6. The two trajectories of the system (4.12). Reprinted from [8]. Copyright (2020) with permission of Springer.

$$
\begin{aligned}
x_{k+1}^1 &= \psi_1(x_k^1, x_k^2, \dots, x_k^n; \mu_1) = r_1 x_k^1(1 - x_k^1) + \mu_1\chi_1(x_k^1, x_k^2, \dots, x_k^n), \\
x_{k+1}^2 &= \psi_2(x_k^1, x_k^2, \dots, x_k^n; \mu_2) = r_2 x_k^2(1 - x_k^2) + \mu_2\chi_2(x_k^1, x_k^2, \dots, x_k^n), \\
&\vdots \\
x_{k+1}^n &= \varphi_n(x_k^1, x_k^2, \dots, x_k^n; \mu_n) = r_n x_k^n(1 - x_k^n) + \mu_n\chi_n(x_k^1, x_k^2, \dots, x_k^n),
\end{aligned}
\tag{4.13}
$$

where μ_i, $i = 1, 2, \dots, n$, are parameters and $\chi = (\chi_1, \chi_2, \dots, \chi_n; \mu_i)$ is a continuous function. Due to the continuity of χ and for r_i, $i = 1, 2, \dots, n$, larger than 4 and sufficiently small μ_i, $i = 1, 2, \dots, n$, one can show that a label set can be constructed using equation (4.13) and thus chaotic behavior in the sense of Poincaré, Li–Yorke and Devaney for an n-dimensional perturbed logistic map would be expected to appear.

4.4 Conservative–progressive growth

The objective of this section is to present a proposal for the construction of fractal curves and fractal functions. Our method can be considered as an alternative to the classical methods. We proceed with the properties of fractal structures found in nature to introduce a mechanism of fractal creation. The generating procedure is differentiated into conservative and progressive ones such that it emulates the evolutionary process in natural fractals such as trees and the human lung. Examples of a random Koch curve and a new fractal function are provided with numerical simulations to illustrate the proposed method. The results of the section are in [4] and are not yet published.

4.4.1 Introduction

Fractal geometry is a scientific field that deals with objects that adopt various degrees of self-similarity. A set that displays self-similarity and repeats the same patterns at every scale is usually called a fractal. Mandelbrot defined a fractal as a set for which the Hausdorff dimension strictly exceeds the topological dimension [36]. The first scientific recognition of fractals appeared as a function when Weierstrass [50] introduced in 1872 a function that is everywhere continuous but nowhere differentiable. Due to the exhibition of the self-similarity property, such a function is now called a fractal function. The idea of self-similar curves received more attention in the work of von Koch. He devised in 1904 a continuous but non-differentiable curve that never intersects itself. Another example of a fractal curve was described by Lévy in 1938, known as the Lévy C curve.

A fractal function can be defined as a mapping from a compact subset in $\mathbb{R}^n$ to $\mathbb{R}$ whose graph has a non-integral Hausdorff dimension [40]. Different examples of fractal functions have been studied in the literature. Singular functions and continuous nowhere differentiable functions were the main types of functions considered [45]. Mandelbrot [35, 40] studied the fractal properties of some fractal function including dimension, similarity and scaling. He introduced the so-called Weierstrass–Mandelbrot fractal which is a reformulation of the Weierstrass function. The idea of fractal functions was developed in [11, 12, 22, 38] where they were introduced as functions whose graphs are the attractors of an iterated function system. Such functions are usually called fractal interpolating functions, where the idea of interpolating a given set of points is applied.

The theory of fractal functions has attracted attention as a useful tool in the applied sciences as well as engineering. The fields of applications include structural mechanics, physics, chemistry, signal processing and decoding, and applied wavelet theory [39]. The classical methods of fractal construction depend on specifying an initiator and a generator, for example figure 4.7 shows the initiator and the generator for the Koch curve. The (conservative) method of the Koch curve construction consists in the consecutive application of the generator which is continuous to arbitrary small scale and this is true for the construction of many fractal curves such as Peano and Hilbert curves. We suggest an alternative method to the conservative methods where, differently, our middle operation works with a discontinuous mediator. We extend the concept of a generator of fractal construction to consist of the *conservator* and the *progressor*, and this is why we call the approach conservative–progressive generation (CPG). The conservative phase can be described as preparation to provide a way to insert new parts into the structure during the progressive stage such that the gaps which are presented by the conservator are filled with the help of the progressor. By consecutive application of the conservator and the progressor to infinitely many steps, one can obtain a fractal curve as a limit position. In more realistic situations a random process can be involved through the progressor action.

4.4.2 The method

Consider an initial line segment F. For the sake of simplification, let us consider a single conservator and double progressor. The conservator action consists of removing a middle open line segment from F to produce two conservative parts

(a) The initiator (b) The generator (c) The Koch curve

Figure 4.7. The classical initiator and the generator for the Koch curve.

denoted as F_1 and F_4 and a gap denoted as F_0. The function of the progressor is to fill the gap by two progressive line segments F^2 and F^3. Here we use superscripts to characterize the action of the progressor. The first step of the fractal construction consists of applying the conservator and the progressor to F to obtain eight lines and four gaps. All parts are denoted by F_i^j such that for conservative parts $i = 1, 4$ and for that progressive parts $j = 2, 3$, whereas the gaps are distinguished by $i = 0$. We repeatedly apply the conservator and the progressor such that at the nth step we denote each element by $F_{i_1 i_2 \ldots i_n}^{j_1 j_2 \cdots j_n}$, $i_k \in \{0, 1, 4\}$, $j_k \in \{1, 2, 3, 4\}$, $k = 1, 2, \ldots, n$. For $i_n = 1, 4$ we have a conservative element, for $j_n = 2, 3$ we have a progressive element and for $i_n = 0$ we have a gap. The sought-after fractal is the set resulting from an infinite step of this procedure. Since each conservative or progressive element is a closed set, the elements with infinite indexes, $F_{i_1 i_2 \ldots i_n \ldots}^{j_1 j_2 \cdots j_n \cdots}$, can be considered as a point. We call the lower sequence, $i_1 i_2 \ldots i_n \ldots$, the *conservative indexes* and the upper sequence, $j_1 j_2 \cdots j_n \ldots$, the *progressive indexes*. The constructed fractal $\mathcal{F}$ is the collection of all such points, i.e.

$$\mathcal{F} = \left\{ F_{i_1 i_2 \ldots i_n \ldots}^{j_1 j_2 \cdots j_n \cdots} \mid i_k \in \{0, 1, 4\},\ j_k \in \{1, 2, 3, 4\},\ k = 1, 2, \ldots \right\}.$$

4.4.3 Koch curve

In this subsection we shall show how to build the Koch curve using the CPG method. Figure 4.8 describes the conservative generator and the progressive generator for the Koch curve. We start with a line segment as the initiator set F, then we iteratively apply the conservator and the progressor. Figure 4.9(a) shows the result of the first step, while part (b) of the figure shows that the second step for the part consists of F_1^1, F_0^1 and F_4^1. By continuing the process to infinity we obtain the classical Koch. The curve is regular since we keep the lengths of the generating lines and orientation of the progressor

(a) The conservator (b) The progressor

Figure 4.8. The conservative generator and the progressive generator for the Koch curve.

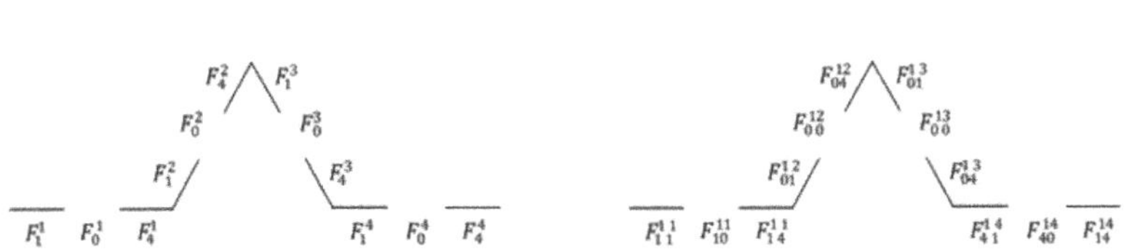

(a) (b)

Figure 4.9. (a) The first step of the construction of the Koch curve. (b) The second step of the construction of the part consisting of F_1^1, F_0^1 and F_4^1.

constant. To generate a random Koch curve we may choose the orientation of the progressor randomly in each step. In this case the obtained curve is a self-similar set and has the same fractal dimension [20]. Another type of random fractal may be introduced by a random choice of the length of the progressive lines. The randomly chosen lines at each step become fixed for the subsequent step. Figure 4.10 shows the first six steps of the construction of such a random Koch curve.

4.4.4 Fractal functions

Similar to the construction of the Koch curve, a fractal function can be created using the CPG method. Let us consider a single conservator and single progressor, as shown in figure 4.11. We start with a curve generated by a function f^0 on an interval $[a, b]$, then we iteratively apply the conservator and the progressor. At each iteration n, the domain interval is divided into 3^{n+1} subintervals denoted by $I_{i_1 i_2 \ldots i_n}^{j_1 j_2 \cdots j_n}$, $i_k \in \{0, 1, 3\}$, $j_k \in \{1, 2, 3\}$, $k = 1, 2, \ldots, n$, such that each element, $F_{i_1 i_2 \ldots i_n}^{j_1 j_2 \cdots j_n}$, of the curve has the same indices as the corresponding subinterval. A gap corresponding to the interval $I_{i_1 i_2 \ldots i_n}^{j_1 j_2 \cdots j_n}$ is filled by the progressor using an appropriately chosen one-to-one function $f^{j_1 j_2 \cdots j_n}$. The conservator action consists of generating the parts $F_{i_1 i_2 \ldots i_n}^{j_1 j_2 \cdots j_n}$, $j_n = 1, 3$, $i_n \neq 0$, which can be defied by

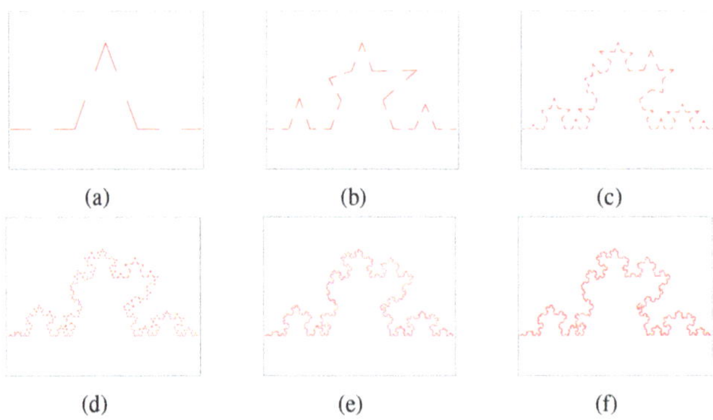

Figure 4.10. The generation of a random Koch curve.

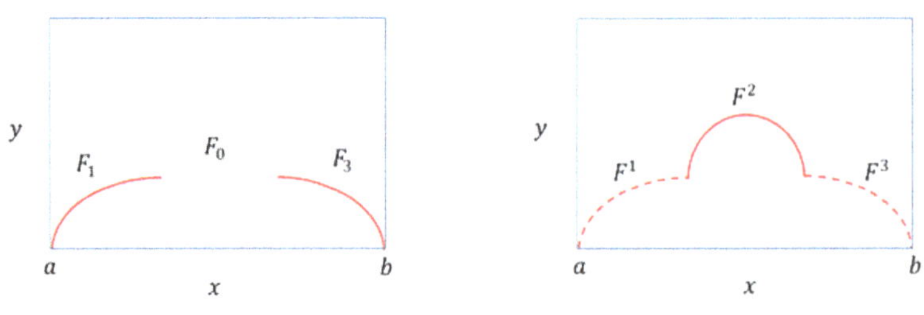

Figure 4.11. The conservative generator and the progressive generator for a fractal function.

$$\left\{(x, y)\colon y = f^{j_1 j_2 \dots j_n}(x),\ \ x \in I^{j_1 j_2 \dots j_n}_{i_1 i_2 \dots i_n},\ \ i_n = 1, 3,\ \ i_n \neq 0\right\}.$$

On the other hand, the progressor action consists of generating the parts $F^{j_1 j_2 \dots j_n}_{i_1 i_2 \dots i_n}$, $j_n = 2$, $i_n \neq 0$, which are defined by

$$\left\{(x, y)\colon y = f^{j_1 j_2 \dots j_n}(x),\ \ x \in I^{j_1 j_2 \dots j_n}_{i_1 i_2 \dots i_n},\ \ j_n = 2,\ \ i_n \neq 0\right\}.$$

The first iteration of the construction of a fractal function graph is depicted in figure 4.12. We continue applying this process and after an infinite number of iterations a graph of a fractal function is obtained. Using the conservative and the progressive indices, each point of the final graph can be represented by $F^{j_1 j_2 \dots j_n \dots}_{i_1 i_2 \dots i_n \dots}$, $i_n \neq 0$. In figure 4.13 we provide a simulation result for an example of fractal function construction. The figure illustrates the first six iterations of the

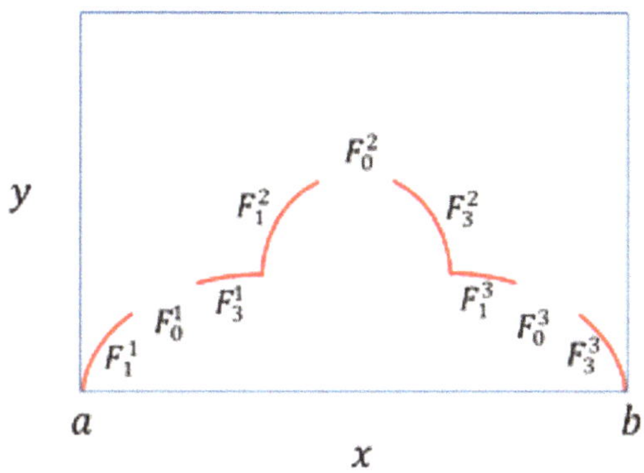

Figure 4.12. The first step of the construction of a fractal function.

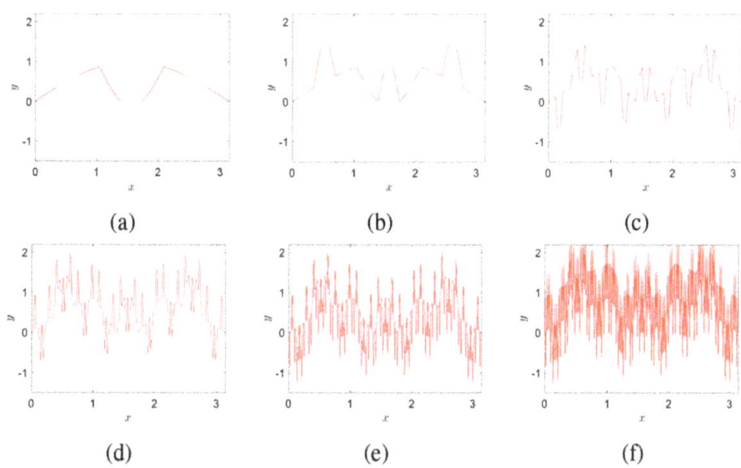

Figure 4.13. Example of a fractal function construction.

Figure 4.14. An illustration of conservative–progressive growth (CPG).

construction. The significance of our approach lies in conceptualizing the growth mechanism of natural fractal structures within a geometric framework for fractal construction. This can be useful to study and analyze complex growth patterns in nature such as tree branching, leaves, fjords, crystals and blood vessel systems [15, 21, 36, 48]. To illustrate this idea we suggest considering part of a fir tree, see figure 4.14. One can decide by comparison of the young and old cones and branches of the tree that there are parts which are lost and there are parts which appear to newly to replace the ones lost during growth. It is important that some elements are saved but are changed. That is, the process of growing is much more complex than just changing quantitatively. This idea is reflected in the way that fractals and fractal functions are constructed, which we call the method of conservative–progressive growth in this section.

4.5 Domain structured dynamics in fractals

There are two different ways to describe an image as a fractal. The first is to consider the Hausdorff dimension of the set. If the number strictly exceeds the topological dimension, then the figure is a fractal. Traditionally fractals are described by self-similarity. This is convenient for our current study. As implementations of abstract self-similarity, we consider several examples of fractals, namely the Sierpinski carpet, Sierpinski gasket, Koch curve and Cantor set. This consists of two main tasks. The first is to indicate abstract self-similarity for the fractals and the second is to ascertain chaos according to the results of the last section.

4.5.1 Sierpinski carpet

Let $\mathcal{S}$ be the Sierpinski carpet constructed in a unit square. In what follows we are going to find a label set for the carpet. We shall denote the abstract set by S. Let us

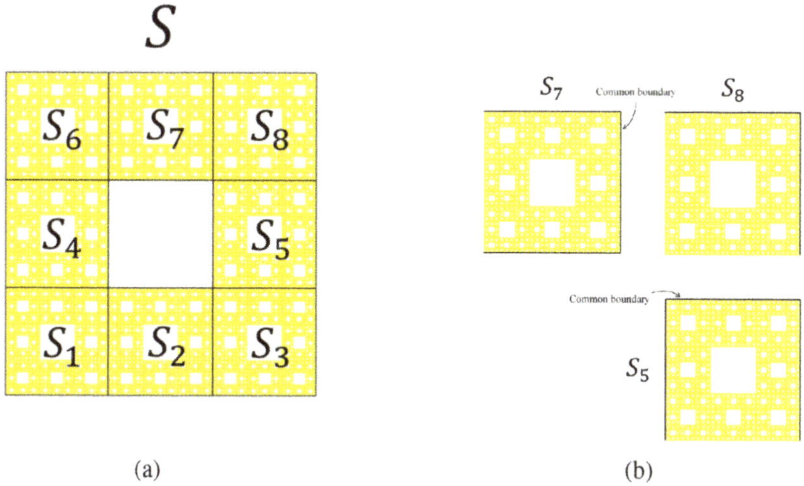

Figure 4.15. (a) The first step of abstract self-similar set construction. (b) The illustration of the boundary agreement. [8], copyright (2020) with permission of Springer. Reprinted from [3], copyright (2021) with permission from AIMS.

start by dividing the carpet into eight subsets and denote them as $S_1, S_2, \ldots S_8$ (see figure 4.15(a)). The subsets will be determined such that any two adjacent subsets have a common horizontal or vertical boundary line. For this reason, we shall use the *boundary agreement* such that: (i) the points of the common boundary of two horizontally adjacent subsets belong to the left one and (ii) the points of the common boundary of two vertically adjacent subsets belong to the lower one. Figure 4.15(b) illustrates the boundary agreement, (i) and (ii), for the subsets, S_5, S_7 and S_8. For clarification the boundaries are shown by black lines and we see that the common boundary points of S_5 and S_8 belong to S_5 and not to S_8 and the common boundary points of S_7 and S_8 belong to S_7 and not to S_8. In the second step each subset S_i, $i = 1, 2, \ldots, 8$, is again subdivided into eight smaller subsets denoted as S_{ij}, $j = 1, 2, \ldots, 8$.

Continuing in the same manner, the subsets of higher order can be determined inductively such that at each nth step we have 8^n subsets notated as $S_{i_1 i_2 \ldots i_n}$, $i_k = 1, 2, \ldots, 8$. Figure 4.16(a) and (b) show, for example, the subsets of S_1 and the subsets of S_{11}, respectively.

To determine the distance between the points of S, we will apply the corresponding Euclidean distance for the set S such that if $\mathbf{x} = (x_1, x_2)$ and $\mathbf{y} = (y_1, y_2)$ are two points in S, then $d(\mathbf{x}, \mathbf{y}) = \sqrt{(x_1 - y_1)^2 + (x_2 - y_2)^2}$. The diameter of a subset at an nth step is equal to $\operatorname{diam}(S_{i_1 i_2 i_3 \ldots i_n}) = \sqrt{2}/3^n$ and, therefore, it diminishes to zero as n tends to infinity and the diameter condition holds. It is easy to check that each point in S has a unique presentation $S_{i_1 i_2 \ldots i_n \ldots}$. Hence, the label set S can be written as

$$S = \left\{ S_{i_1 i_2 \ldots i_n \ldots} : i_k = 1, 2, \ldots 8, \ k \in \mathbb{N} \right\}.$$

The separation condition of degree 1 is satisfied and the separation constant is

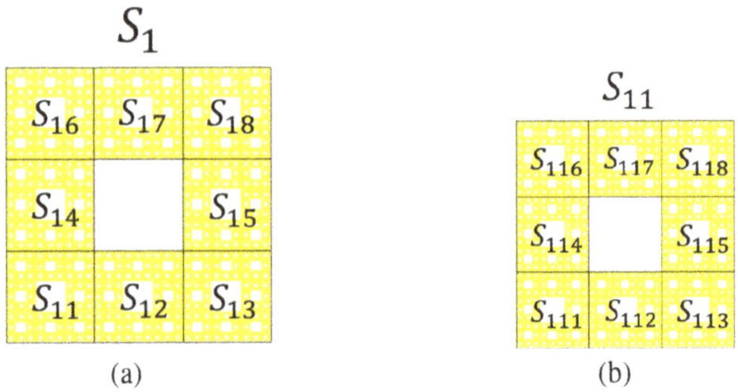

Figure 4.16. Examples of the second and the third order subsets of the Sierpinski carpet. [8], copyright (2020) with permission of Springer. Reprinted from [3], copyright (2021) with permission from AIMS.

$$\varepsilon_0 = \left\{\min\{d(S_i, S_j)\}: S_i, S_j \text{ are disjoint}, i, j = 1, 2, \dots, 8\right\} = \frac{1}{3}.$$

Let us now define the similarity map by

$$\varphi(S_{i_1 i_2 i_3 \dots}) = S_{i_2 i_3 \dots}.$$

Thus we have shown that the triple (S, d, φ) is a domain structured dynamics with the separation and diameter conditions. In view of theorems 4.1.2, 4.1.1 and 4.1.3, the similarity map ψ is chaotic in the sense of Poincaré, Li–Yorke and Devaney.

4.5.2 A chaotic trajectory

In this subsection we provide a geometric realization of the similarity map on the Sierpinski carpet and see how the map can be useful for indexing the subsets of a self-similar set and visualizing the trajectories of its points. A chaotic trajectory is seen as expected in the last subsection. In [7] we adopt the idea of Fatou–Julia iteration (sometimes called the ETA) and develop a scheme for constructing the Sierpinski carpet. The scheme is based on the iterations of the modified planar tent map

$$T(x) = \begin{cases} 3\,[x(\text{mod}1)] & x \leqslant \frac{1}{2} \text{ or } x > 1, \\ 3(1 - x) & \frac{1}{2} < x \leqslant 1, \end{cases}$$

$$T(y) = \begin{cases} 3\,[y(\text{mod}1)] & y \leqslant \frac{1}{2} \text{ or } y > 1, \\ 3(1 - y) & \frac{1}{2} < y \leqslant 1. \end{cases}$$

Depending on this one can construct a map T_i: $S \to S$ such that the set S is invariant,

$$T_i(x) = \begin{cases} 3x & 0 \leqslant x \leqslant \frac{1}{3}, \\ 3x - 1 & \frac{1}{3} < x \leqslant \frac{1}{2}, \\ 2 - 3(1 - x) & \frac{1}{2} < x < \frac{2}{3}, \\ 3(1 - x) & \frac{2}{3} \leqslant x \leqslant 1, \end{cases}$$

$$T_i(y) = \begin{cases} 3y & 0 \leqslant y \leqslant \frac{1}{3}, \\ 3y - 1 & \frac{1}{3} < y \leqslant \frac{1}{2}, \\ 2 - 3(1 - y) & \frac{1}{2} < y < \frac{2}{3}, \\ 3(1 - y) & \frac{2}{3} \leqslant y \leqslant 1. \end{cases} \tag{4.14}$$

This map is equivalent to the similarity map φ defined above, therefore the trajectory of a point $\mathbf{x} \in S$ can be visualized using the map (4.14). Figure 4.17 shows an example of the trajectory for the center point $\mathbf{x}$ of the subset

$$S_{277\,311\,373\,132\,771\,824\,315\,158\,224\,617\,847\,648\,526\,563\,584\,625\,456\,271\,254\,233\,172\,162\,44}.$$

The points of the trajectory are considered as the centers of the subsets which are determined by $T_i^k(\mathbf{x})$, $k = 0, 1, 2, \ldots 68$.

Figure 4.17. The trajectory of a point under the similarity map. [8], copyright (2020) with permission of Springer. Reprinted from [3], copyright (2021) with permission from AIMS.

4.5.3 Sierpinski gasket

To construct an abstract label set on the basis of the Sierpinski gasket, let us consider the Sierpinski gasket generated in a unit equilateral triangle. We first divide the gasket into three smaller parts to be the first order subsets and denote them by G_1, G_2 and G_3, as shown in figure 4.18(a). By glancing at the figure, one can see that every two subsets share only a single point as a common boundary. For this reason we consider the following boundary agreement: the common boundary point of every couple of adjacent subsets belongs either to the left one or to the lower one. Applying the agreement, the subsets G_1, G_2 and G_3 become disjoint subsets of the desired abstract self-similar set G such that $G = \cup_{i=1}^{3} G_i$. Second, each subset, G_i, $i = 1, 2, 3$, is again subdivided into three subsets and we denote them as G_{ij}, $i, j = 1, 2, 3$ (see figure 4.18(b)). Taking into account the boundary agreement we repeat the same procedure such that at each nth step we denote the resultant subsets by $G_{i_1 i_2 \ldots i_n}$, $i_k = 1, 2, 3$.

The subsets of the Sierpinski gasket described above have an inverse relationship with the construction-step variable, n, $\operatorname{diam}(G_{i_1 i_2 i_3 \ldots i_n}) = 1/2^n$, from which one can verify the validity of the diameter condition.

Following the arguments of the abstract self-similarity one can deduce that a point in G can be uniquely represented by $G_{i_1 i_2 \ldots i_n \ldots}$ and the label set G can be expressed as

$$G = \left\{ G_{i_1 i_2 \ldots i_n \ldots} \colon i_k = 1, 2, 3, \ k \in \mathbb{N} \right\}.$$

Contrary to the Sierpinski carpet case the separation constant ε_0 for the Sierpinski gasket cannot be evaluated through the first order subsets G_i, $i = 1, 2, 3$. This is why we considered the minimum of the distance between any two disjoint subsets of the second order, that is,

$$\varepsilon_0 = \{\min\{d(G_{i_1 i_2}, G_{j_1 j_2})\} \colon G_{i_1 i_2}, G_{j_1 j_2} \text{ are disjoint}, \ i_n = 1, 2, 3\} = \frac{\sqrt{3}}{8},$$

where d is the usual Euclidean distance. Thus one can see that the separation condition is valid.

The similarity map acting on the Sierpinski gasket, G, can be defined by $\varphi(G_{i_1 i_2 i_3 \ldots}) = G_{i_2 i_3 \ldots}$. Consequently, the triple (G, d, φ) is a domain structured

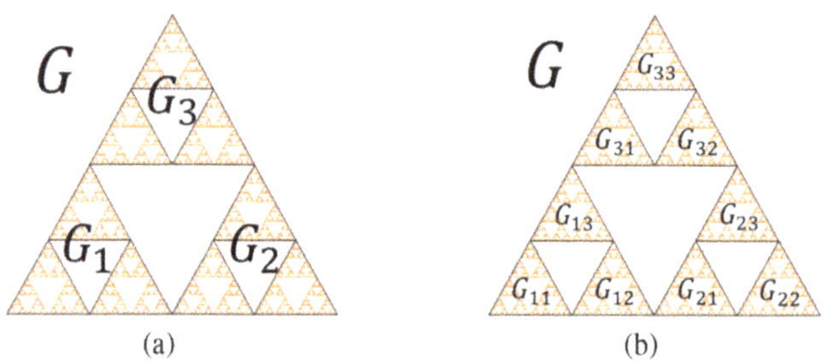

Figure 4.18. The construction of abstract self-similar sets corresponding to the Sierpinski gasket. [8], copyright (2020) with permission of Springer. Reprinted from [3], copyright (2021) with permission from AIMS.

dynamics with diameter and separation conditions and φ is chaotic in the sense of Poincaré, Li–Yorke and Devaney.

The same idea can be extended to the fractals associated with Pascal's triangles. It is well known that Pascal's triangle in mod 2 creates the classical Sierpinski gasket. Different fractals associated with Pascal's triangles in different moduli can be considered as abstract similar sets and it can also be proved that the similarity map defined on these sets possesses chaos.

4.5.4 Koch snowflake

Let us consider the Koch curve, $\mathcal{K}$, constructed from an initial unit line segment. To identify an abstract self-similar set K corresponding to the Koch snowflake we start by dividing the set into four equal parts (subsets) and denoting them as K_1, K_2, K_3 and K_4, as shown in figure 4.19(a). Since the Koch curve is a connected set each two adjacent subsets share a single point as a common boundary. Let us denote the end points of each subset K_i, $i = 2, 3$ by a_i and a_{i+1}. Figure 4.19(a) illustrates these points and for clarification the points are shown by black thick dots. It is seen in the figure that K_1 and K_2 share the point a_2, K_2 and K_3 share the point a_3, and K_3 and K_4 share the point a_4. In the second step each subset K_i, $i = 1, 2, 3, 4$, is again subdivided into four subsets K_{ij}, $j = 1, 2, 3, 4$. Figure 4.19(b) and (c) illustrate the second step for the subsets K_1 and K_2, respectively. Again here we see that each two adjacent subsets share a single boundary point. We continue in the same way such that at each step, n, every set $K_{i_1i_2\ldots i_{n-1}}$, $i_k = 1, 2, 3, 4$, is divided into four subsets $K_{i_1i_2\ldots i_{n-1}i_n}$, $i_n = 1, 2, 3, 4$, and denote the end points of each $K_{i_1i_2\ldots i_n}$, $i_n = 2, 3$, by $a_{i_1i_2\ldots i_n}$ and $a_{i_1i_2\ldots i_n+1}$.

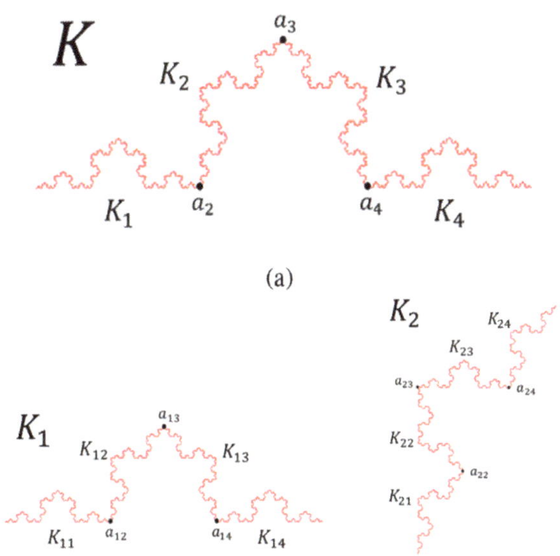

Figure 4.19. The construction of an abstract self-similar set corresponding to the Koch curve. [8], copyright (2020) with permission of Springer. Reprinted from [3], copyright (2021) with permission from AIMS.

As in the previous cases, to determine the abstract self-similar set we need to consider the following boundary agreement: for each pair of adjacent subsets $K_{i_1i_2\ldots i_{n-1}j}$ and $K_{i_1i_2\ldots i_{n-1}j+1}$, the common boundary point $a_{i_1i_2\ldots i_{n-1}j+1}$ belongs to $K_{i_1i_2\ldots i_{n-1}j+1}$. This condition means that the common boundary point a_2 shown in figure 4.19(a), for example, belongs to K_2 not to K_1 and the common boundary point a_{23} shown in figure 4.19(c) belongs to K_{23} not to K_{22}.

By applying the boundary agreement to all subsets at each step, we have fully described the disjoint subsets of the proposed abstract self-similar set for the Koch snowflake. From the construction of the Koch curve and by using the usual Euclidean distance, one can deduce that the distance between the end points of each subset $K_{i_1i_2\ldots i_n}$ is $\frac{1}{3^n}$, which clearly represents the diameter of the subset. Therefore, the diameter condition holds. A point in K can be represented by $K_{i_1i_2\ldots i_n\ldots}$, so that

$$K = \left\{K_{i_1i_2\ldots i_n\ldots}\colon i_k = 1, 2, 3, 4, \ k \in \mathbb{N}\right\}.$$

The separation condition is also valid with degree 1, since for any K_i, $i = 1, 2, 3, 4$, one can find K_j, $j = 1, 2, 3, 4$, $j \neq i$, such that they are separated from each other by a distance of not less than ε_0. The separation constant, ε_0, can be defined by

$$\varepsilon_0 = \min\{d(K_1, K_3), d(K_1, K_4), d(K_2, K_4)\} = \frac{\sqrt{7}}{9}.$$

The similarity map for the abstract fractals of Koch snowflake is given by $\varphi(K_{i_1i_2i_3\ldots}) = K_{i_2i_3\ldots}$ and, thus, we have shown that the triple (K, d, φ) is a chaotic domain structured dynamics.

4.5.5 Cantor set

A perfect example of chaos in fractals is the Cantor set. As we previously mentioned, the chaoticity in the Cantor set is determined by finding a topological conjugacy with the symbolic dynamics. To show that the Cantor set is not an exception to our approach for chaos, we shall establish an abstract self-similar set corresponding to the Cantor set. Let us consider the middle third Cantor set, $\mathcal{C}$, initiated from a unit line segment. Denote the corresponding label set C. The first step consists of dividing the set into two subsets and denoting them by C_1 and C_2 (see figure 4.20(a)). In the

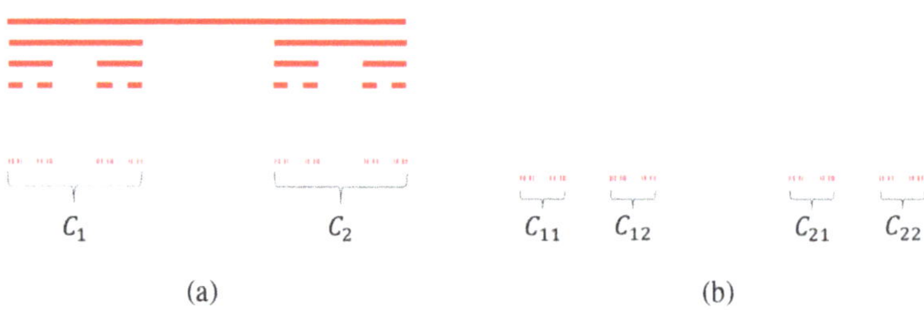

Figure 4.20. The first and the second order subsets for the Cantor set. [8], copyright (2020) with permission of Springer. Reprinted from [3], copyright (2021) with permission from AIMS.

second step each of C_1 and C_2 is subdivided into two subsets, as shown in figure 4.20(b). These subsets are denoted by C_{11}, C_{12}, C_{21} and C_{22}. In every subsequent step, we repeat the same procedure for each subset resulting from the preceding step. We denote the resultant subsets at each nth step by $C_{i_1i_2\ldots i_n}$, $i_k = 1, 2$.

In the Cantor set case we do not need any boundary agreement since all subsets are disjoint. Considering the usual Euclidean distance the diameter of a subset, $C_{i_1i_2\ldots i_n}$, is $1/3^n$. Therefore the diameter condition holds. The points in C are represented by $C_{i_1i_2\ldots i_n\ldots}$. Hence the label set is defined by

$$C = \left\{ C_{i_1i_2\ldots i_n\ldots}\colon i_k = 1, 2, \ k \in \mathbb{N} \right\}.$$

From the construction the separation condition is clearly valid and the constant ε_0 is defined by the distance between the subsets C_1 and C_2 so that $\varepsilon_0 = 1/3$.

The similarity map is defined by $\varphi(C_{i_1i_2i_3\ldots}) = C_{i_2i_3\ldots}$ and the triple (C, d, φ) is a domain structured dynamics. Theorems 4.1.2, 4.1.1 and 4.1.3 are also applicable for this case.

In connection with the above examples of chaos, we remark that the Sierpinski carpet and Koch curve indicate that a non-continuous map can have a domain which is a connected set, while for continuous maps the domains of chaos are usually disconnected. Examples of disconnected chaotic domains are the Cantor set for the logistic map [17], the modified Sierpinski triangle with exceptions in [10], the set associated with Smale's horseshoe map [52] and the Poincaré section of the Lorenz attractor [37].

4.6 Notes

We are strongly confident that the method of domain structured dynamics is a very effective instrument for theoretical analysis, first of all for sophisticated dynamics. There are many various models where the structures can appear. One can apply the method in the following three ways: *structural dynamization, chaos verification* and *stochastic processes.* These three are connected to what we call sophisticated or complex or irregular dynamics and are detailed in the introductory part of the chapter. Another source of interesting problems is the way in which structuring can be performed. For the moment we are able to suggest two tools for this. The first one is that based on similarity and the other is the Fatou–Julia iterations. They are relative to each other. However, the iterations method may be considered as the primary one. In other words, if a structuring is obtained by similarity then it can be done by iterations, but not vice versa. However, generally speaking they have to be equivalent. This is the time to explain in more detail what we accept as structuring by similarity and what by iterations. The first method is naturally popular in the determination of fractals by self-similarity in the traditional manner, for example, the Sierpinski carpet or the Cantor set. The approach is suitable for non-fractal sets as well. This is seen, for example, in the chaos arrangement for finite-dimensional cubes in this chapter. The Fatou–Julia method of the abstract similarity set construction is easy in our case. One can consider a bounded area and then collect all points which leave the set after n-iterations and call the collection envelope for the

n-dimensional rectangle. One can finalize the structuring by considering the intersection of the envelopes. To structure by this method we need the initial first order partition. The best example is the Cantor middle set, but this time determined by the logistic map iterations. We hope that one can establish in the same way the abstract similarity chaos for the Julia set and other sophisticated figures. Moreover, the algorithm will be of use for abstract constructions and, consequently, for dynamical equations. One can expect that there is an amazing relationship between what we call the conservative–progressive growth of fractals in this chapter and the results by Turing on morphogenesis [47].

References

[1] Addabbo T, Fort A, Rocchi S and Vignoli V 2011 Digitized chaos for pseudo-random number generation in cryptography *Chaos-based Cryptography: Theory, Algorithms and Applications* ed L Kocarev and S Lian (Berlin: Springer) pp 67–97

[2] Akhmet M and Alejaily E M 2019 Domain-structured chaos in a Hopfield neural network *Int. J. Bifurcat. Chaos* **29** 1950205

[3] Akhmet M and Alejaily E M 2021 Abstract similarity, chaos and fractals *Discrete Contin. Dyn. Syst. Ser.* B **26** 2479–97

[4] Akhmet M and Alejaily E M 2020 *CPG Method for Fractal Curves and Functions* (submitted)

[5] Akhmet M and Fen M O 2016 Poincaré chaos and unpredictable functions *Commun. Nonlinear Sci. Numer. Simul.* **48** 85–94

[6] Akhmet M and Fen M O 2016 Unpredictable points and chaos *Commun. Nonlinear Sci. Numer. Simul.* **40** 1–5

[7] Akhmet M, Fen M O and Alejaily E M 2018 Dynamics motivated by Sierpinski fractals arXiv: 1811.07122

[8] Akhmet M, Fen M O and Alejaily E M 2020 *Dynamics with Chaos and Fractals* (Berlin: Springer)

[9] Avila A and Moreira C G 2003 Bifurcations of unimodal maps *Dynamical Systems* pp 1–22

[10] Barnsley M 1988 *Fractals Everywhere* (New York: Academic)

[11] Barnsley M F and Demko S 1985 Iterated function systems and the global construction of fractals *Proc. R. Soc.* A **399** 243–75

[12] Barnsley M F, Elton J, Hardin D P and Massopust P R 1989 Hidden variable fractal interpolation functions *SIAM J. Math. Anal.* **20** 1218–42

[13] Bolotin Y, Tur A and Yanovsky V 2009 *Chaos: Concept, Control and Constructive Use* (Berlin: Springer)

[14] Brown R 1928 A brief account of microscopical observations made in the months of June, July and August, 1927, on the particles contained in the pollen of plants; and on the general existence of active molecules in organic and inorganic bodies *Philos. Mag. Ann. Philos. New Ser.* **4** 161–73

[15] Carbone A, Gromov M and Prusinkiewicz P 2000 *Pattern Formation in Biology, Vision and Dynamics* (Singapore: World Scientific)

[16] Chen G and Huang Y 2011 *Chaotic Maps: Dynamics, Fractals and Rapid Fluctuations, Synthesis Lectures on Mathematics and Statistics* (San Rafael, CA: Morgan and Claypool)

[17] Devaney R L 1989 *An Introduction to Chaotic Dynamical Systems* (Canada; USA: Addison-Wesley)

[18] Değirmenci N and Koçak Ş 2010 Chaos in product maps *Turk. J. Math.* **34** 593–600
[19] Feigenbaum M J 1980 Universal behavior in nonlinear systems *Los Alamos Sci.* **1** 4–27
[20] Feldman D P 2012 *Chaos and Fractals: An Elementary Introduction* (Oxford: Oxford University Press)
[21] Hahn H K, Georg M and Peitgen H-O 2005 Fractal aspects of three-dimensional vascular constructive optimization *Fractals in Biology and Medicine* (Berlin: Birkhäuser) 55–66
[22] Hardin D P and Massopust P R 1986 The capacity for a class of functions *Commun. Math. Phys.* **105** 455–60
[23] Hilborn R C 2000 *Chaos and Nonlinear Dynamics: An Introduction for Scientists and Engineers* (New York: Oxford University Press)
[24] Hunt B R and Kaloshin V Y 2010 Prevalence *Handbook of Dynamical Systems* ed H W Broer, V Y Hasselblatt and F Takens (Amsterdam: Elsevier) pp 43–87
[25] Ivancevic V G and Ivancevic T T 2007 *High-Dimensional Chaotic and Attractor Systems* (Dordrecht: Springer)
[26] Julia G 1918 Mémoire suv i'itération des fonctions rationelles *J. Math. Pures Appl.* **8** 47–245
[27] Keener J P 1980 Chaotic behavior in piecewise continuous difference equations *Trans. Am. Math. Soc.* **261** 589–604
[28] Kellert S H 1994 *In the Wake of Chaos: Unpredictable Order in Dynamical Systems* (Chicago, IL: University of Chicago Press)
[29] Korsch H J, Jodl H J and Hartmann T 2008 *Chaos: A Program Collection for the PC* (Berlin: Springer)
[30] Layek G C 2015 *An Introduction to Dynamical Systems and Chaos* (India: Springer)
[31] Li R and Zhou X 2013 A note on chaos in product maps *Turk. J. Math* **37** 665–75
[32] Li T Y and Yorke J A 1975 Period three implies chaos *Am. Math. Mon.* **82** 985–92
[33] Lorenz E N 1984 The local structure of chaotic attractor in four dimensions *Physica* D **13** 90–104
[34] Lynch S 2017 *Dynamical Systems with Applications Using Mathematica* 2nd edn (Cham: Springer)
[35] Mandelbrot B B 1977 *Fractals: Form, Chance, and Dimension* (San Francisco, CA: Freeman)
[36] Mandelbrot B B 1983 *The Fractal Geometry of Nature* (New York: Freeman)
[37] Masoller C, Schifino A C and Sicardi R L 1995 Characterization of strange attractors of Lorenz model of general circulation of the atmosphere *Chaos Soliton. Fract.* **6** 357–66
[38] Massopust P R 1987 Dynamical systems, fractal functions and dimension *Topol. Proc.* **12** 93–110
[39] Massopust P R 1997 Fractal functions and their applications *Chaos Solitons Fractals* **8** 171–90
[40] Massopust P R 2004 Fractal functions and wavelets: examples of multiscale theory *Abstract and Applied Analysis, Proc. of the Int. Conf., Hanoi, Vietnam* (Singapore: World Scientific Press) pp 225–47
[41] Mehlhorn K 1984 *Data Structures and Algorithms* vol 3 (Berlin: Springer)
[42] Osikawa M and Oono Y 1981 Chaos in C^0-endomorphism of interval *Publ. RIMS* **17** 165–77
[43] Sander E and Yorke J A 2011 Period-doubling cascades galore *Ergodic Theory Dyn. Syst.* **31** 1249–67
[44] Schöll E and Schuster H G (ed) 2008 *Handbook of Chaos Control* (Weinheim: Wiley)
[45] Singer P and Zajdler P 1999 Self-affine fractal functions and wavelet series *J. Math. Anal. Appl.* **240** 518–51

[46] Skiadas C H and Skiadas C (ed) 2016 *Handbook of Applications of Chaos Theory* (Boca Raton, FL: Chapman and Hall)

[47] Turing A 1952 The chemical basis of morphogenesis *Philos. Trans. R. Soc. Lond.* B **237** 32–72

[48] Vicsek T 1992 *Fractal Growth Phenomena* (Singapore: World Scientific)

[49] Wang M C and Uhlenbeck G E 1945 On the theory of Brownian motion II *Rev. Mod. Phys.* **17** 323–42

[50] Weierstrass K 1895 Über continuirliche Functionen eines reellen Arguments, die für keinen Werth des Letzteren einen Bestimmten Differentialquotienten besitzen *K Weierstrass, Mathematische Werke* (Berlin: Mayer and Miiller)

[51] Wiggins S 1988 *Global Bifurcation and Chaos, Analytical Methods* (New York: Springer)

[52] Zeraoulia E and Sprott J C 2011 *Robust Chaos and Its Applications* (Singapore: World Scientific)

Part III

Random processes

Chapter 5

Randomness with similarity dynamics

The results of this chapter are not surprising if one takes into account the following simple implication. Loosely speaking, the theory of stochastic processes consists of a list of *necessities* which have been found for the dynamics with chance. Mathematical chaos is composed of ingredients which confirm *necessarily* the dynamics. This is why we may look for the ingredients of chaos in random dynamics. If they occur and are proved there for certain one can speak of the presence of chaos in stochastic processes. To be clear, we do not speak of determinism in processes with chance but of non-trivial ingredients which are common for deterministic chaos and random dynamics. This time we have chosen unpredictable sequences, the main ingredient of Poincaré chaos, for the discussion. The choice is proved by advanced characteristics of the concept, namely Poisson stability and unpredictability. The sequences are proposed for certain occurrence in random experiments of Bernoulli schemes and Markov chains. We apply the method of domain-structured chaos for a dynamics with orbits whirling among infinitely many modules. Thus, the conservative conditions of the existence of chaos are weakened to model irregular dynamics. The research proves that Poincaré chaos is of exceptional use in the analysis of stochastic processes. To confirm Poincaré chaos in a random process with probability of one, we can simulate an unpredictable function which is defined through realization of a Bernoulli scheme. We have contributed to the theory of probability by shaping the characteristics in the notion of unpredictable strings. New laws of large strings are formulated and verified for Bernoulli schemes. Thus, we have proposed new necessary events for probabilistic analysis. We hope that they are present in the processes sufficiently intensively to be applied, for example, in stochastic differential equations theory.

5.1 The basics of abstract similarity

In this section we will introduce the basic characteristics of abstract similarity dynamics and, more specifically, abstract similarity chaos. Correspondingly, we

doi:10.1088/978-0-7503-3507-2ch5

should start with the description of abstract similarity sets, which are cornerstones of the dynamics. They will be used for abstract similarity dynamics through labeling in the state space of random processes. Since labeling is utilized we will say that a state space is structured or a domain is structured. The labeling can be performed with multiplicity one and then the domain is an uncountable set. This time we apply infinite multiplicity since the state spaces are finite. The multiplicity is convenient for random processes, which are the focus of this chapter.

5.1.1 Abstract similarity

In this section *abstract similarity* as a concept and as a mapping is considered for the abstract label set. It is obvious that similarity can be defined only by a mapping and metrical properties are also desirable. We specify the label set as a self-similar set and a collection of points in a metric space where it can be considered as a union of infinite shrinking sets with notation that allows us to introduce the dynamics in the set. Then one can prove chaos as the most important result of the current research. For this purpose a specific map over the invariant abstract similarity set is defined. The map acts as the identity if the whole set taken as an argument for the map. This feature is equivalent to the property of self-similarity in ordinary fractals.

The first of our suggestions is a collection

$$\mathcal{F} = \{\mathcal{F}_{i_1 i_2 \ldots i_n \ldots}\colon i_k = 1, 2, \ldots , m, \;\; k = 1, 2, \ldots\}, \tag{5.1}$$

where m is a natural number and members of the set are called *labels*. The set $\mathcal{F}$ is called the *abstract similarity set*. Another name, the *label set*, is also sensible. It will be very useful for the development of the theory of abstract similarity dynamics regardless of applications. We suggest that one can develop the theory of abstract similarity dynamics similar to that constructed for symbolic dynamics in [40].

For dynamical requirements we shall use the following subsets of the label set, *n-dimensional rectangles*,

$$\mathcal{F}_{i_1 i_2 \ldots i_n} = \bigcup_{j_k = 1, 2, \ldots, m} \mathcal{F}_{i_1 i_2 \ldots i_n j_1 j_2 \ldots}, \tag{5.2}$$

where indices $i_1, i_2, \ldots , i_n$ are fixed.

One can easily verify that

$$\mathcal{F} \supseteq \mathcal{F}_{i_1} \supseteq \mathcal{F}_{i_1 i_2} \supseteq \ldots \supseteq \mathcal{F}_{i_1 i_2 \ldots i_n} \supseteq \mathcal{F}_{i_1 i_2 \ldots i_n i_{n+1}} \ldots, \;\; i_k = 1, 2, \ldots , m, \;\; k = 1, 2, \ldots.$$

That is, the rectangles form a nested sequence.

Next the algorithm of application by labeling is briefly described. Consider a compact metric space (F, d) with the distance d. In the current chapter we suggest that the set F is finite, but generally it can be countable and even uncountable. We will say that an abstract similarity set $\mathcal{F}$ is a label set for F, if each element of the set F is labeled by at least a single member of the set $\mathcal{F}$ and each element of the set $\mathcal{F}$ presents a member from F. Thus we do not suggest that uniqueness is required for the relation. That is, if $\mathcal{F}_{i_1 i_2 \ldots i_n \ldots}$ and $\mathcal{F}_{j_1 j_2 \ldots jn, \ldots}$ present the same element f of F it is not necessary that $i_n = j_n$ for all $n = 1, 2, \ldots$. Moreover, we keep the distance d for the set $\mathcal{F}$ considering $d(\mathcal{F}_{i_1 i_2 \ldots i_n \ldots}, \mathcal{F}_{j_1 j_2 \ldots j_n, \ldots}) = d(f_1, f_2)$, if $\mathcal{F}_{i_1 i_2 \ldots i_n \ldots}$ and $\mathcal{F}_{j_1 j_2 \ldots j_n, \ldots}$ are labels of elements f_1 and f_2 of the set F, respectively, such that $d(\mathcal{F}_{i_1 i_2 \ldots i_n \ldots}, \mathcal{F}_{j_1 j_2 \ldots j_n, \ldots}) = 0$ for different labels of the same point in F. If there is uniqueness in the labeling the couple $(\mathcal{F}, d)$ is also a metric space.

Let us consider the label set, $\mathcal{F}$, and determine the map φ: $\mathcal{F} \to \mathcal{F}$ such that

$$\varphi(\mathcal{F}_{i_1 i_2 \ldots i_n \ldots}) = \mathcal{F}_{i_2 i_3 \ldots i_n \ldots}. \tag{5.3}$$

Considering iterations of the map, applied to an n-dimensional rectangle, one can verify that

$$\varphi^n(\mathcal{F}_{i_1 i_2 \ldots i_n}) = \mathcal{F}, \tag{5.4}$$

where $i_k = 1, 2, \ldots, m$, $k = 1, 2, \ldots, n$. The relations (5.3) and (5.4) give us a reason to call φ the *similarity map* and the number n the *order of similarity*. It is reasonable to call φ the *abstract similarity map* as it is in the abstract space $(\mathcal{F}, d)$. Moreover, it is clear that in this book we can accept the terms *label set* and *abstract similarity set* as synonyms. Finally, if a compact metric space (F, d) is given without any dynamics, we call the triple $(\mathcal{F}, d, \varphi)$ the *abstract similarity dynamics* in the space (F, d) under the condition that $\mathcal{F}$ is a label set for F. We will say that for rectangles $\mathcal{F}_{i_1 i_2 \ldots i_n}$ the *diameter condition* is valid if

$$\sup_{i_k=1,2,\ldots,m} \operatorname{diam} \mathcal{F}_{i_1 i_2 \ldots i_n} \to 0 \text{ as } n \to \infty, \tag{5.5}$$

where $\operatorname{diam} A = \sup\{d(\mathbf{x}, \mathbf{y}) : \mathbf{x}, \mathbf{y} \in A\}$ for a set A in $\mathcal{F}$.

5.1.2 Chaotic similarity

To prove chaos for the abstract similarity dynamics we assume in this section the *separation condition*. Define the distance between two nonempty bounded sets A and B in $\mathcal{F}$ by $d(A, B) = \inf\{d(\mathbf{x}, \mathbf{y}) : \mathbf{x} \in A, \mathbf{y} \in B\}$. The abstract similarity set $\mathcal{F}$ satisfies the separation condition of degree n if there exists a positive number ε_0 such that for arbitrary string $i_1 i_2 \ldots i_n$ one can find a string $j_1 j_2 \ldots j_n$ which satisfies

$$d\left(\mathcal{F}_{i_1 i_2 \ldots i_n}, \mathcal{F}_{j_1 j_2 \ldots j_n}\right) \geqslant \varepsilon_0. \tag{5.6}$$

We call ε_0 the *separation constant*.

For the Poincaré chaos the unpredictable motion is utilized to distinguish the chaotic behavior instead of the periodic motions in the Devaney and Li–Yorke types. The existence of infinitely many unpredictable Poisson stable trajectories that lie in a compact set meet all the requirements of chaos. Based on this, chaos appears in the dynamics on the quasi-minimal set which is the closure of a Poisson stable trajectory. Therefore the Poincaré chaos is referred to as the dynamics on the quasi-minimal set of trajectory initiated from an unpredictable point. The next theorem shows that the Poincaré chaos is valid for the similarity dynamics.

Let us give the definition of the unpredictability in the sense of the dynamics $(\mathcal{F}, d)$.,

Definition 5.1.1. [10] *A point $\mathcal{F}^*_{i_1 i_2 \ldots}$ from the set $\mathcal{F}$ is unpredictable if there exist a positive number ε_0 and sequences t_m, s_m, of natural numbers both of which diverge to infinity such that $\varphi^{t_m}(\mathcal{F}_{i_1 i_2 \ldots i_k \ldots}) \to \mathcal{F}_{i_1 i_2 \ldots i_k \ldots}$ as $m \to \infty$ and $d(\varphi^{t_m + s_m}(\mathcal{F}_{i_1 i_2 \ldots i_k \ldots}), \varphi^{t_m}(\mathcal{F}_{i_1 i_2 \ldots i_k \ldots})) \geqslant \varepsilon_0$ for each natural number m.*

Theorem 5.1.1. [8] *If the separation and diameter conditions are valid, then the abstract similarity map possesses Poincaré chaos.*

The proof of theorem 5.1.1 can be found in chapter 3.

In the following theorem, see chapter 3, it is proved that the similarity map φ possesses the three ingredients of Devaney chaos, namely the density of periodic points, transitivity and sensitivity. A point $\mathcal{F}_{i_1 i_2 i_3 \ldots} \in \mathcal{F}$ is periodic with period n if its index consists of endless repetitions of a block of n terms.

Theorem 5.1.2. [8] *If the separation and diameter conditions hold, then the abstract similarity dynamics is chaotic in the sense of Devaney.*

In addition to the Devaney and Poincaré chaos, it can be shown that the Li–Yorke chaos also takes place in the dynamics of the map φ. The proof of the following theorem is similar to that of theorem 6.35 in [22] for the shift map defined on the space of symbolic sequences.

Theorem 5.1.3. [8] *The abstract similarity map is Li–Yorke chaotic if the separation and diameter conditions hold.*

We shall call the behavior for $(\mathcal{F}, \varphi, d)$ the *abstract similarity chaos*. Thus if the diameter and separation conditions are valid then the dynamics has abstract similarity chaos, which assumes all three types of chaos: Poincaré, Li–Yorke and Devaney.

Now we explain how the abstract similarity chaos can be applied for chaos indication in models. Assume a dynamics (F, ψ, d) is given. Let $\mathcal{F}$ be a label set for F. Assume that the separation and diameter conditions for the space $(\mathcal{F}, d)$ are valid and by theorems 5.1.2, 5.1.1 and 5.1.3 the dynamics $(\mathcal{F}, \varphi, d)$ has abstract similarity chaos, i.e. it is Poincaré, Li–Yorke and Devaney chaotic. Assume that the mappings ψ and φ agree such that if $\mathcal{F}_{i_1 i_2 \ldots}$ is a label for an element f of F, then $\mathcal{F}_{i_2 i_3 \ldots}$ is a label for $\psi(f)$. One can easily see that, if the agreement is valid, then the dynamics (F, ψ, d) is also Poincaré, Li–Yorke and Devaney chaotic. It is sensible to call the dynamics of ψ *domain-structured chaos*. Assume also the case that a metric space (F, d) is not provided a dynamics. Then we say that the metric space is domain-structured chaotic if there exists an abstract label set $\mathcal{F}$ for F such that $(\mathcal{F}, \varphi, d)$ is abstract similarly chaotic. The last notion is useful for fractals analysis. The assigned dynamics in space (F, d) is called domain-structured chaos. Let us consider the following example, which is a good starting point for the discussion of randomness.

Example 5.1.1. *Consider the finite metric space* (F, d) *of two elements f and g. Assign to f and g infinitely many labels* $\mathcal{F}_{1i_2\ldots}$ *and* $\mathcal{F}_{2i_2\ldots}$, *respectively, such that* $i_k = 1, 2$ *for all* $k = 2, 3, \ldots$. *It is clear that the separation and diameter conditions are valid for the dynamics* $(\mathcal{F}, \varphi, d)$. *Now introduce in the space* (F, d) *the mapping* ψ, *which agrees with* φ. *Thus we have obtained domain-structured chaotic dynamics* (F, ψ, d). *The dynamics will be useful to discuss random processes, in particular Bernoulli schemes.*

5.2 Unpredictability in Bernoulli schemes

Fix the alphabet $1, 2, \ldots, N$ and consider the set $\Omega = \{1, 2, \ldots, N\}^{\mathbb{N}}$, where $\mathbb{N}$ is the set of all natural numbers, as a sample set. Elements ω of Ω are infinite sequences $(\omega_1, \omega_2, \ldots)$, where ω_i, $i \in \mathbb{N}$, are members of the alphabet. Additionally, we consider finite strings $(\omega_1, \ldots, \omega_n)$, $n \in \mathbb{N}$. They are members of the cylindrical sets $\Omega^n = \{1, 2, \ldots, N\}^n$. Assume that the probabilities of all members of the alphabet are positive and the sum of probabilities is equal to unity. Determine a family of random variables, $X(n, \omega)$: $\mathbb{N} \to S$, where $S = \{s_1, \ldots, s_m\}$, $m \in \mathbb{N}$, is a finite set of elements. Let a distance d be assigned for S such that the couple (S, d) is a metric space and denote $\varepsilon_0 = \min\{d(s_k, s_l)\colon k, l = 1, 2, \ldots, m, k \neq l\}$. It is clear that $\varepsilon_0 > 0$. We shall call the sequences $\{X(k, \omega)\}$, $k \in \mathbb{N}$, and $\{X(k, \omega)\}_{k=1}^{n}$, $\omega \in \Omega$, the infinite and finite realizations of the Bernoulli scheme, respectively. The notion of the realization for the Bernoulli scheme is one of the basic ideas of this section. The infinite realization is a sequence $\{a_k\}_k$, $k \in \mathbb{N}$, and the finite realization is a sequence $\{a_k\}$, $1 \leqslant k \leqslant n$, $n \in \mathbb{N}$, with $a_k \in S$, $k \in \mathbb{N}$. They are orbits of the dynamics, which we know as the Bernoulli scheme. In the case $N = 2$ the dynamics is said to be the Bernoulli process [21, 54].

Assign for each member s_i, $i = 1, \ldots, m$, of the state space of the Bernoulli scheme infinitely many labels $\mathcal{F}_{ii_2\ldots}$, where $i_j = 1, \ldots, m$, $j = 2, \ldots$. Thus one can consider the label set $\mathcal{F} = \{F_{i_1i_2\ldots}, i_j = 1, \ldots, m, j = 1, \ldots\}$. It is easy to verify that the diameter and separation conditions are valid for the set $\mathcal{F}$. That is, the set is a chaotic structure and the abstract similarity chaos is valid for the dynamics $(\mathcal{F}, \varphi, d)$, where φ is the abstract similarity map. In light of the last section's results, this means that Poincaré, Li–Yorke and Devaney chaos are present for the triple. In what follows we will pay attention only to Poincaré chaos and more precisely to the unpredictable sequence. This notion provides strict evidence for the Poincaré chaos and vice versa [10].

Definition 5.2.1. [10, 11] *A bounded sequence* κ_i, $i \geqslant 1$, *in* $\mathbb{R}^p$ *is called unpredictable if there exist a positive number* ε_0 *and sequences* ζ_n, η_n, $n \geqslant 1$, *of positive integers both of which diverge to infinity such that* $\|\kappa_{i+\zeta_n} - \kappa_i\| \to 0$ *as* $n \to \infty$ *for each i in bounded intervals of integers and* $\|\kappa_{\zeta_n+\eta_n} - \kappa_{\eta_n}\| \geqslant \varepsilon_0$ *for each natural number n.*

For a finite metric space (S, d) the last definition has the following form.

Definition 5.2.2. *A bounded sequence $s_i \in S$, $i \geqslant 0$, is called unpredictable if there exist a positive number ε_0 and sequences ζ_n, η_n, $n \geqslant 0$, of positive integers both of which diverge to infinity such that $s_{i+\zeta_n} = s_i$ for each bounded interval of integers, if n is sufficiently large, and $d(s_{\zeta_n+\eta_n}, s_{\eta_n}) \geqslant \varepsilon_0$ for each natural number n.*

The closure of the unpredictable orbit of the abstract similarity chaos is the quasi-minimal set [11]. It contains an uncountable set of unpredictable trajectories. Fix an unpredictable orbit of the map φ. It is also an unpredictable infinite realization of the Bernoulli scheme. Let us fix an unpredictable realization. Due to the definition 5.2.2 and theorem 5.1.1, the following assertion is valid.

Theorem 5.2.1. *Each finite realization of the Bernoulli scheme coincides with an arc of the unpredictable realization. That is, the unpredictable realization happens in each experiment of the scheme and is a certain event.*

Thus one has to recognize that each experiment in the Bernoulli scheme generates an arc of the unpredictable orbit. It is in full accordance with the principle of the ergodic theory [57] that a single trajectory proves behavior of the whole dynamics and all other trajectories. In other words, the random process is a reproduction of the chaos with probability one.

This is the moment to explain why we have chosen the unpredictable trajectory and, hence, the Poincaré chaos for the discussion instead of the other types of chaos. One may consider the transitive orbit of the Devaney chaos. It is, as well as the unpredictable orbit, dense in the domain of chaos. The choice of the unpredictable sequence is preferable since it admits more constructive characteristics than the transitive orbit. They are the sequence of moments of time, when the values of a motion recurrently visit the initial state as well those which guarantee the divergence. The results are useful for research and applications of random processes, such as coin tossing, dice rolling, traffic lights, tetrahedron dice rolling and five city entrance.

5.3 Random unpredictable functions

A special type of the Poisson stable orbit, the unpredictable trajectory, which leads to Poincaré chaos was introduced in [11]. The subsequent papers [10, 12] were concerned with the unpredictable functions of continuous and discrete arguments. This is why, to confirm the Poincaré chaos in a random process with probability one, we can simulate an unpredictable function, which is defined through realization of a Bernoulli scheme.

Definition 5.3.1. [12] *A uniformly continuous and bounded function $v: \mathbb{R} \to \mathbb{R}^p$ is unpredictable if there exist positive numbers ε_0, δ and sequences t_n, u_n all of which diverge to infinity such that $v(t + t_n) \to v(t)$ as $n \to \infty$ uniformly on compact subsets of $\mathbb{R}$ and $\|v(t + t_n) - v(t)\| \geqslant \varepsilon_0$ for each $t \in [u_n - \delta, u_n + \delta]$ and $n \in \mathbb{N}$.*

In [12] it is shown that each simulation of the equation

$$x' = -2x + \pi(t) \tag{5.7}$$

asymptotically approaches the solution of the equation (5.7), which is an unpredictable function in the sense of the definition 5.3.1 if the perturbation $\pi(t)$ is based on an unpredictable sequence in the way described below. In what follows we perform the simulation using a random unpredictable sequence.

Consider the Bernoulli process, that is the discrete-time random process with $m = 2$, $s_1 = 1$, $s_2 = 2$ and $d(s_1, s_2) = |1 - 2| = 1$, and take into account the corresponding dynamics of the similarity map such that $\mathcal{S}_{1i_2\ldots,i_k,\ldots}$ is assigned for 1 and $\mathcal{S}_{2i_2\ldots}$ is assigned for 2, where $i_j = 1, 2, j = 2, \ldots$. We assume that $d(\mathcal{S}_{i_1i_2\ldots}, \mathcal{S}_{j_1j_2\ldots}) = d(s_{i_1}, s_{j_1})$ if $i_1, j_1 = 1, 2,$ to obtain the chaotic structure $\mathcal{S} = \{\mathcal{S}_{i_1i_2\ldots}, i_j = 1, 2, j = 1, \ldots\}$. From theorem 5.2.1 each finite realization is an arc of an unpredictable sequence. Let us fix a finite realization i_k, $i_k = 1, 2$, $k \geqslant 1$, of the Bernoulli process. Consider the piece-wise constant function, $\pi(t)$, defined on the positive real semi-axis through the equation $\pi(t) = i_k$ for $t \in \left[\frac{k}{100}, \frac{k+1}{100}\right)$, $k \in \mathbb{Z}$. This function satisfies the requirements of [12].

Consider the function

$$\Theta(t) = \int_{-\infty}^{t} e^{-2(t-s)}\pi(s)ds. \tag{5.8}$$

It is worth noting that $\Theta(t)$ is the unique, bounded on the whole real axis, globally exponentially stable solution of the differential equation (5.7).

One can confirm that the function $\Theta(t)$ is uniformly continuous since its derivative is bounded. Because the sequence $\{i_k\}$ is unpredictable, there exist a positive number ε_0 and sequences $\{\zeta_n\}$, $\{\eta_n\}$ both of which diverge to infinity such that $|i_{k+\zeta_n} - i_k| \to 0$ as $n \to \infty$ for each k in bounded intervals of integers and $|i_{\zeta_n+\eta_n} - i_{\eta_n}| \geqslant \varepsilon_0$ for each $n \in \mathbb{N}$. Fix an arbitrary positive number ε and arbitrary real numbers α, β with $\beta > \alpha$. Let N be a sufficiently large natural number satisfying

$$N \geqslant \ln\left(\frac{2}{3\varepsilon}\right).$$

There exists a natural number n_0 such that for each $n \geqslant n_0$ the inequality

$$|i_{k+\zeta_n} - i_k| < \frac{\varepsilon}{2}$$

holds for $k = 100(\lfloor\alpha\rfloor - N), 100(\lfloor\alpha\rfloor - N) + 1, \ldots, 100\lfloor\beta\rfloor$, where $\lfloor\alpha\rfloor$ and $\lfloor\beta\rfloor$, respectively, denote the largest integers which are not greater than α and β. Accordingly, if $n \geqslant n_0$, then we have

$$|\pi(t + \zeta_n/100) - \pi(t)| < \frac{\varepsilon}{2}, \tag{5.9}$$

for $t \in [\lfloor\alpha\rfloor - N, \lfloor\beta\rfloor + 1)$.

The last inequality implies that $|\Theta(t + \zeta_n/100) - \Theta(t)| < \varepsilon$ for $t \in [\lfloor \alpha \rfloor, \lfloor \beta \rfloor + 1]$. Hence $|\Theta(t + \zeta_n/100) - \Theta(t)| \to 0$ as $n \to \infty$ uniformly on the interval $[\alpha, \beta]$.

On the other hand, one can confirm for each $n \in \mathbb{N}$ that $|\pi(t + \zeta_n/100) - \pi(t)| \geqslant \varepsilon_0$ for $t \in [\eta_n/100, (\eta_n + 1)/100)$. For fixed $n \in \mathbb{N}$, using the equation

$$\begin{aligned}\Theta(t + \zeta_n/100) - \Theta(t) = {} & \Theta(\zeta_n/100 + \eta_n/100) - \Theta(\eta_n/100) \\ & - 2\int_{\eta_n/100}^{t} (\Theta(s + \zeta_n/100) - \Theta(s))ds \\ & + \int_{\eta_n/100}^{t} (\pi(s + \zeta_n/100) - \pi(s))ds,\end{aligned}$$

we obtain that

$$\begin{aligned}& |\Theta(\zeta_n/100 + \eta_n/100 + 1/100) - \Theta(\eta_n/100 + 1/100)| \\ & \geqslant \left| \int_{\eta_n/100}^{\eta_n/100+1/100} (\pi(s + \zeta_n/100) - \pi(s))ds \right| \\ & - |\Theta(\zeta_n/100 + \eta_n/100) - \Theta(\eta_n/100)| \\ & - 2\left| \int_{\eta_n/100}^{\eta_n/100+1/100} (\Theta(s + \zeta_n/100) - \Theta(s))ds \right|.\end{aligned}$$

Therefore, one can verify that

$$\sup_{t \in [\eta_n/100, (\eta_n+1)/100]} |\Theta(t + \zeta_n/100) - \Theta(t)| \geqslant \frac{\varepsilon_0}{202}.$$

Thus there exists a sequence $\{u_n\}$ with $\eta_n/100 \leqslant u_n \leqslant (\eta_n + 1)/100$, $n \in \mathbb{N}$, such that

$$|\ \Theta(u_n + \zeta_n/100) - \Theta(u_n)| \geqslant \frac{\varepsilon_0}{202}.$$

For $t \in [u_n - \delta, u_n + \delta]$, where $\delta = \varepsilon_0/2424$, we have that

$$\begin{aligned}|\Theta(t + \zeta_n/100) - \Theta(t)| \geqslant {} & |\Theta(u_n + \zeta_n/100) - \Theta(u_n)| \\ & - \frac{5}{2}\left| \int_{u_n}^{t} |\ \Theta(s + \zeta_n/100) - \Theta(s)|\ ds \right| \\ & - \left| \int_{u_n}^{t} |\ \pi(s + \zeta_n/100) - \pi(s)|\ ds \right| \geqslant \frac{\varepsilon_0}{404}.\end{aligned}$$

It is clear that $u_n \to \infty$ as $n \to \infty$. Thus the function $\Theta(t)$ is unpredictable. This is the first case when the deterministic unpredictable function has been determined by a random process. To visualize the function we shall apply the asymptotic stability of the unpredictable solution. This means that one can simulate any solution of the equation and it will approach ultimately the unpredictable function. Accordingly, we depict in figure 5.1 the solution $\Psi(t)$, $\Psi(0) = 0.5$ of the equation (5.7), which approximates the unpredictable solution $\Theta(t)$.

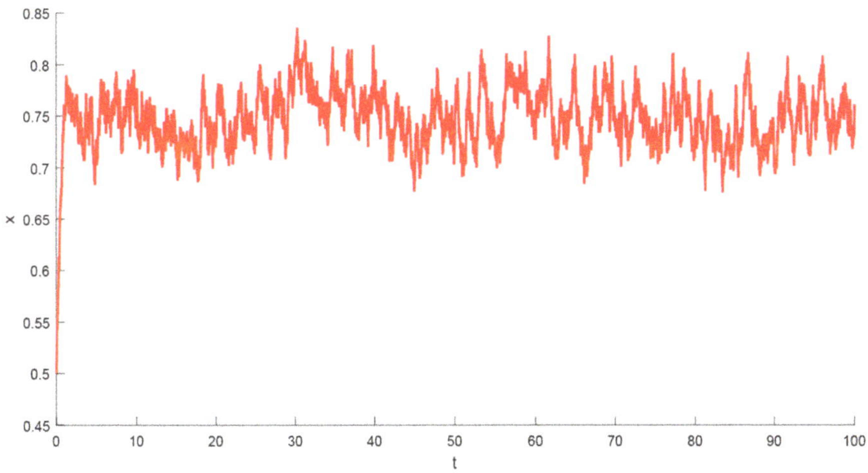

Figure 5.1. The solution $\Psi(t)$, which approximates asymptotically $\Theta(t)$.

5.4 Unpredictable strings

The Bernoulli process, and the Bernoulli scheme in general, are among the most attractive and basic concepts in probability theory, statistics, random processes and dynamical processes related to chaos [1, 16, 19, 24, 30, 33–36, 45–49, 54, 58]. In the current section we suggest considering, in addition to the traditional probabilistic events, a new one, which is called an unpredictable string. It will provide interesting opportunities for the extension of the theories as well as exploration of useful deterministic features for stochastic dynamics. The notion of infinite sequences with unpredictable strings is introduced. Numerical simulations of the Bernoulli process are performed to demonstrate that the realizations are unpredictable. They confirm that specific properties for the random dynamics are valid, namely the first and second laws of large (unpredictable) strings, which are also discussed theoretically. In addition, a Matlab algorithm is provided to verify sequences with inductively increasing lengths of unpredictable strings. The results of section 5.2 are useful for comparison as they are more general in some respects.

5.4.1 Basic dynamics of strings

Next we introduce the main concept of this section, unpredictable strings, and utilize it to determine unpredictable sequences.

Let a_i, $i = 0, 1, 2, \ldots$, be an infinite sequence of symbols.

Definition 5.4.1. [14] *A finite array* $(a_s, a_{s+1}, \ldots, a_{s+k})$, *where* s *and* k *are positive integers, is said to be an unpredictable string of length* k *if* $a_i = a_{s+i}$, *for* $i = 0, 1, 2, \ldots, k-1$, *and* $a_k \neq a_{s+k}$.

The diagram in figure 5.2 illustrates the definition.

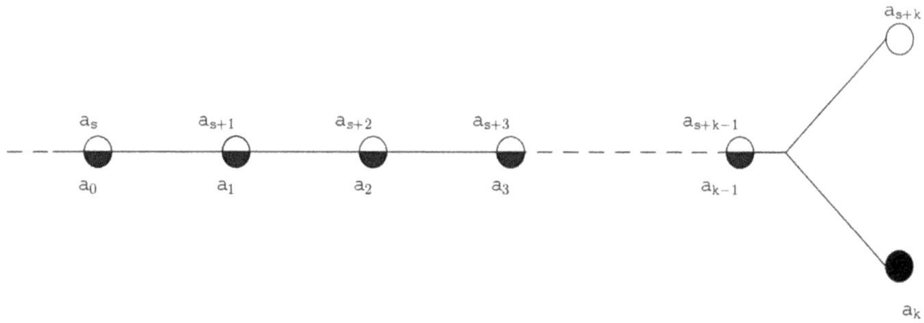

Figure 5.2. An illustration of the unpredictable string of length k.

Definition 5.4.2. [14] *The sequence a_i, $i \geqslant 0$, is unpredictable if it admits unpredictable strings with arbitrary large lengths.*

Definition 5.4.3. [14] *The sequence a_i, $i \geqslant 0$, is unpredictable if there exist sequences ζ_n, η_n of positive integers both of which diverge to infinity such that $a_{\zeta_n+l} = a_l$, $l = 0, 1, 2, \ldots, \eta_n - 1$ and $a_{\zeta_n+\eta_n} \neq a_{\eta_n}$, for each $n \in \mathbb{N}$.*

Theorem 5.4.1. [14] *The definitions* 5.4.2 *and* 5.4.3 *are equivalent.*

Proof. Let sequence a_i be unpredictable in the sense of definition 5.4.3. Then the finite arrays $(a_{\zeta_n}, a_{\zeta_n+1}, \ldots, a_{\zeta_n+\eta_n})$ are unpredictable strings of length η_n, for each natural n. Thus the sequence admits unpredictable strings with arbitrarily large lengths.

Conversely, let a_i be a sequence that admits unpredictable strings of arbitrary large lengths, i.e. there is a sequence i_n, $n = 1, 2, 3, \ldots$, such that the finite arrays $(a_{i_n}, a_{i_n+1}, \ldots, a_{i_n+k})$ are unpredictable strings. By setting $\zeta_n = i_n$ and $\eta_n = i_n + k$ we deduce that the sequence a_i is unpredictable in light of definition 5.4.3. □

Fix a positive integer k and denote by S_k the sets of all indices s such that the strings $(a_s, a_{s+1}, \ldots, a_{s+k})$ are unpredictable within the sequence a_i, $i = 1, 2, \ldots$, which is not necessarily unpredictable.

Theorem 5.4.2. [14] *The sets S_l and S_q do not intersect if $l < q$.*

Proof. Assume, in contrast, that sets S_l and S_q admit a common element s. Then we have that $a_l \neq a_{s+l}$ if $s \in S_l$ and $a_l = a_{s+l}$ if $s \in S_q$. This contradiction completes the proof. □

Theorem 5.4.3. [14] *Assume that a_i, $i \geqslant 0$, is an unpredictable sequence. Then each a_j with positive j is the first element of an unpredictable string, if $a_j = a_0$.*

Proof. Assume the opposite. Then one can show that the sequence α is a periodic one. That is not an unpredictable sequence. □

Example 5.4.1. [14] *To have a better impression of the unpredictable strings, let us consider the graph of the piece-wise constant function, $H(t)$, which values on intervals $[i/10, (i+1)/10)$, $i = 0, 1, \dots, 199$, are assigned randomly 1 or -1 with equal probability 1/2. The two unpredictable strings as a result of the Bernoulli process are present, in red, in figure 5.3(a). The second, with a length of 0.7 units, is placed between coordinates 14 and 16, shown in figure 5.3(c), while its corresponding initial arc is shown in figure 5.3(b).*

5.4.2 Strings in coin tossing

We will scrutinize a realization of the Bernoulli process as a sequence consisting of the digits 1 and 0 with positive probabilities. That is, the model for coin tossing.

First we provide an algorithm for the indication of unpredictable strings in realizations of a Bernoulli scheme on a finite number of complex vectors $v_1, v_2, \dots, v_r$.

Let us set $a_0 = \text{random}(\{v_1, v_2, \dots, v_r\})$ and $a_1 = \text{random}(\{v_1, v_2, \dots, v_r\})$. Then for increasing $k = 1, 2, 3, \dots$ we define $a_{m(k)+j} = a_j$, for $j < k$, and $a_{m(k)+j} = \text{random}(\{v_1, v_2, \dots, v_r\} - a_j)$, for $j = k$, where $m(k+1) = m(k) + k$ with $m(1) = 2$.

The sequence

$$(0, 1, 0, 0, 0, 1, 1, 0, 1, 0, 1, 0, 1, 0, 0, 1, 0, 1, 0, 0, 0, 0, 0, 1, 0, 0, 0, 1, 0, \dots)$$

is a result of the application of the algorithm with $v_1 = 0$ and $v_2 = 1$.

Let us introduce several characteristics that are of use for the analyses of finite realizations of the Bernoulli scheme. For a fixed natural number m, consider a finite realization, a_i, $i = 0, 1, \dots, m$. Denote by $K(m)$ the largest length of unpredictable strings in the array. For every k between 1 and $K(m)$, denote by q_k the number of k-length unpredictable strings within the array, by ξ_k the largest index such that $(a_{\xi_k}, a_{\xi_{k+1}}, \dots, a_{\xi_k+k})$ is an unpredictable string within the array and by $N(m)$ the

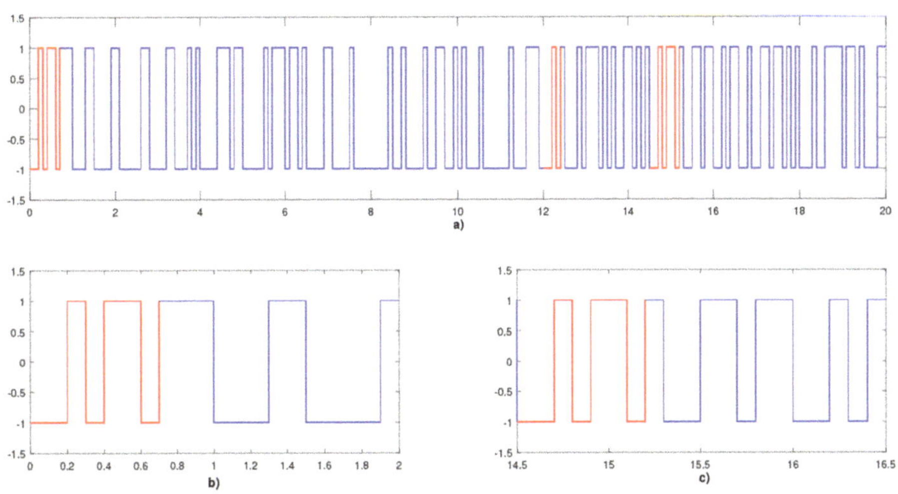

Figure 5.3. The graph of the function $H(t)$, which illustrates the appearance of unpredictable strings.

Table 5.1. The values $K(n)$, $\xi_{K(n)}$ and $N(n)/n$ for the finite realization.

n	$K(n)$	$\xi_{K(n)}$	$N(n)/n$
50	10	20	0.72
200	10	20	0.58
500	10	228	0.586
2000	14	1008	0.596
10 000	14	3469	0.603 1
20 000	18	19 206	0.599 5
100 000	21	74 683	0.601 4
500 000	21	401 088	0.600 3
800 000	21	663 684	0.600 1
900 000	28	874 766	0.568 6

number of all unpredictable strings, which have a non-empty intersection with the array.

Now we provide statistical results on the realization, which are obtained through Matlab simulations for the Bernoulli process with probability $p = 0.6$ and $m = 9 \times 10^5$. We have evaluated the values of $K(n)$, $\xi_{K(n)}$ and $N(n)/n$ for each n from 1 to m. Ten samples of the simulations are provided in table 5.1. According to the full data obtained in the simulations, the realization can be considered as part of an unpredictable sequence, since there are unpredictable strings with increasing lengths. Moreover, $N(n)/n \approx p$ if n is large.

5.4.3 Laws of large strings

In this section we consider a discrete-time random process $\mathbf{X}(n)$ with the finite state space of r different symbols $s_1, s_2, \dots, s_r$. The function admits values s_i with positive probabilities p_i, $i = 1, 2, \dots, r$, the sum of which is equal to the unit. A realization α of the process is the sequence a_i, $i = 1, 2, \dots$, and a finite realization α_m is the array a_i, $i = 1, 2, \dots, m$. We claim that stochastic processes with discrete-time and finite state spaces satisfy the following theorem.

Theorem 5.4.4. [14] *(The first law of large strings.) The discrete-time random process $X(n)$ with the finite state space admits an uncountable set of realizations, which are unpredictable sequences in the sense of definition* 5.4.2.

Proof. Let us consider the space Σ_r of infinite sequences of a finite set of symbols $s_1, s_2, \dots, s_r$, with the metric

$$d(\xi, \zeta) = \Sigma_{k=0}^{\infty} \frac{|\xi_k - \zeta_k|}{2^k}, \tag{5.10}$$

where $\xi = (\xi_0\xi_1\xi_2\dots)$, $\zeta = (\zeta_0\zeta_1\zeta_2\dots)$. The Bernoulli shift σ on Σ_r is defined as $\sigma(\xi_0\xi_1\xi_2\dots) = (\xi_1\xi_2\xi_3\dots)$. The map is continuous and Σ_r is a compact metric space [58].

It is clear that the set of all realizations of the random dynamics $X(n)$ coincides with the set of all sequences of the symbolic dynamics on Σ_r. According to the result in [10], see also example 2.1.4, the symbolic dynamics admits an unpredictable point, i^*, a sequence from the set Σ_r. There is an uncountable set of unpredictable points, which are unpredictable sequences in the sense of definition 5.4.2. □

It is important that each realization is in the closure for the unpredictable orbit. The density is considered in the shift dynamics sense. The property of the metric implies that each arc of any sequence in the space coincides with some arc of the unpredictable sequence.

Denote by $n(m)$ the number of elements which are equal to a_0 in a finite string. The limit $E[a_0] = \lim_{m\to\infty} n(m)/m$ is said to be the expected value such that $E[a_0] = p_i$, if $a_0 = s_i$, $i = 1, \dots, r$ [21].

Theorem 5.4.3 implies the equality $N(m) = n(m)$, where $N(m)$ is the number of unpredictable strings, which intersect the array. Hence the following proposition is correct, which can be useful for applications.

Theorem 5.4.5. [14] *If a realization α is an unpredictable sequence, then the relation*

$$\lim_{m\to\infty} \frac{N(m)}{m} = E[a_0], \tag{5.11}$$

is valid.

Theorem 5.4.6. [14] *(The second law of large strings.) If the discrete-time random process $X(n)$ admits a finite state space, then the relation*

$$\lim_{m\to\infty} P\left(\left|\frac{N(m)}{m} - E[a_0]\right| < \varepsilon\right) = 1, \tag{5.12}$$

holds for any $\varepsilon > 0$.

Proof. Due to theorem 5.4.1 the Bernoulli scheme is Poincaré chaotic. Consequently, there is an unpredictable realization which is dense in the set of all other realizations with a topology of uniform convergence on each bounded interval of integers. Because of the finite space set, for each finite realization there exists an arc of the unpredictable sequence, which coincides with the realization. Now, taking into account the time-uniform property we can conclude that each of the finite realizations is, in fact, an arc of an unpredictable orbit and, consequently, from the last theorem the proof is complete. □

The last result is in full accordance with the theorem 5.2.1, which says that any finite realization is an arc of a fixed unpredictable realization common for all experiments.

5.5 Abstract similarity chaos in Markov chains

Here we formalized realizations of Markov chains as conveniently constructed sequences and explain why the random dynamics admits the unpredictability, considering Poincaré chaos. An illustrative example with proper numerical simulations is provided.

5.5.1 Introduction and preliminaries

The first goal of the Markov research [39] was to show that the random processes of dependent events may behave as for independent events. Thus relatively simple models were invented, which are most effective for many applications. It is impossible to underestimate the role of the Markov chains in the development of random dynamics theory and its applications. For example, the ergodic theorem was strictly proved for the first time under the conditions. There are many observations that the chains are relative to symbolic dynamics and correspondingly to the Bernoulli scheme. The significant step for understanding this was performed by Ornstein in [48], who verified that B automorphisms, such as subshifts of finite type and Markov shifts, Anosov flows and Sinai's billiards, ergodic automorphisms of the n-torus [51] and the continued fraction transforms, are in fact isomorphic. Researchers have looked in both directions, for chaos in random dynamics as well as for stochastic features in deterministic motions [15, 18, 19]. Thus the problem of chaos in Markov chains, which is discussed in this chapter, is part of the more general project.

The unpredictable orbit [11] as a single isolated motion, presenting Poincaré chaos [10], and identified in the chains as a certain event, is the ultimate point of the comprehension of the results of this chapter. In other words, the main result of this paper is the existence of a realization, an unpredictable sequence, with the finite state space, the closure of which, in the topology of convergence on bounded intervals, is the set of all infinite realizations. One must specify that the convergence, in the circumstances, is the coincidence of the unpredictable realization with each finite realization. This implies that each finite simulation of the dynamics is an arc of the unpredictable realization. In other words, one can confirm that *unpredictability is a certain event for the Markov chain.* Let us recall that *an event is certain if it occurs during every performance of an experiment.*

The main task of the current research is to find signs of unpredictability in Markov chains and then to show that the presence is maximal in the sense of probability. The method of domain-structured chaos originated in [7, 8] is at the basis of our study. It has been successfully applied for chaos indication in fractals [8], neural networks [7], and for Bernoulli processes. This time, we demonstrate that each finite realization of a Markov chain is an arc of a fixed unpredictable orbit common to all finite realizations.

One can apply finite [15, 19] or countable [20] partitioning to find randomness in deterministic motions. In contrast, we utilize the uncountable partitioning [8, 13] to establish strong relations of the random processes with Poincaré chaos. We formalize chaos generation through specially structured sets, subduing them to the similarity map. This provides us with the comprehension that chaotic behavior is proper to the Bernoulli scheme dynamics and then to various random and deterministic processes in discrete and continuous time [5, 6]. Currently, we are on the next step of the exploration and show how chaos can be recognized in Markov chains. Time-homogeneous chains with a finite state space and memory are under discussion. The approach is applied such that the dynamics can be investigated, focusing not on ergodicity, but on a single motion description such that an individual geometry is better understood. Thus, we show that an isolated realization of a Markov chain behaves in time identically with a properly chosen realization of a Bernoulli scheme, and they are both identical in dynamics with the path of the similarity map in a correspondingly chosen space.

The closure of the unpredictable realization is said to be the quasi-minimal set [53]. It contains an uncountable set of unpredictable realizations. We have proved that *a quasi-minimal set as the union of all infinite realizations of the Markov chain is a certain event*. In other words, each experiment of the process produces an orbit of the quasi-minimal set. This is another formulation of the main result.

Our research confirms that the Bernoulli scheme, Markov chains as well as abstract hyperbolic dynamics [5] all have the same type of chaos. It is significant that the Poincaré chaos is proper for the motions. This provides new opportunities, particularly for stochastic processes. We suppose that the research can be complemented with similar analysis for other Bernoulli automorphisms considered by Ornstein [48] as well with extension of the results for the majority of stochastic processes, if proper structured domains are constructed.

5.5.2 Markov chains

A Markov chain is a stochastic model which describes a sequence of possible events such that the probability of each event depends only on the state obtained in the previous one [30, 34]. There are many applications of the Markov chains as statistical models of real-world processes such as studying queues or lines of customers arriving at an airport, currency exchange rates, cruise control systems in motor vehicles and animal population dynamics [41].

Let a discrete-time stochastic process X_n, $n \geqslant 0$, on a countable set S be given. That is, a collection of random variables defined on a probability space (Ω, F, P), where P is a probability measure on a family of events F in an event-space Ω. The set S is the state space of the process and the value $X_n \in S$ is the state of the process at time n. The Markov chain is a stochastic process such that the Markov property $P\{X_{n+1} = s_j|X_0, \dots, X_n\} = P\{X_{n+1} = s_j|X_n\}$ is true for all $s_i, s_j \in S$ and $n \geqslant 0$, and $P\{X_{n+1} = s_j|X_n = s_i\} = p_{ij}$, where p_{ij} is the transition probability that the chain jumps from state i to state j. The property says that, at any time n, the next state X_{n+1} is conditionally independent of the past $X_0, \dots, X_{n-1}$ given the present state X_n. More precisely, that the transition probabilities do not depend on the time parameter n. That is, the chain is time-homogeneous. If the transition probabilities were functions of time, the process would be a non-time-homogeneous Markov chain.

Consider a finite state space $S = \{s_1, \dots, s_m\}$, where m is a natural number, not smaller than two, and a metric d for the space. Denote by $p_{ij} = p_j(s_i)$ the Markov probability for s_j such that $\Sigma_{j=1}^{m} p_{ij} = 1$ for all $i = 1, \dots, m$. In what follows we shall investigate the problem of chaos considering that all the probabilities p_{ij}, $i, j = 1, \dots, m$, are positive.

If f_{ij} is the event which consists of two elementary events s_i and s_j occurring successively, then an infinite realization of the Markov chain X_n, $n \geqslant 0$, can be formalized as the infinite sequence $f_{i_1 j_1} f_{i_2 j_2} \dots f_{i_n j_n} \dots$, with $i_k = j_{k-1}$ for all $k = 2, 3, \dots$. We have that $p(f_{ij}) = p_{ij}$. The formalization does not provide an advantage if one considers the chains without memory, but it makes the discussion of the processes with memory easier in what follows.

Present the last sequence as the element $\mathcal{F}_{i_1 i_2 \dots}$, $i_k = 1, 2, \dots, m$, $k = 1, 2, \dots$, of the space $\mathcal{F}$ with the metric $\delta(\mathcal{F}_{i_1 i_2 \dots}, \mathcal{F}_{l_1 l_2 \dots}) = \Sigma_{k=1}^{\infty} d(s_{i_k}, s_{l_k})/2^k$.

Next, we will consider Markov chains with memory of a non-zero length, in addition to the zero length memory. We start with a length equal to two such that the element f_{ijk} presents three elementary events s_i, s_j and s_k occurring successively, and the probability for s_k is equal to $p_{ijk} = p_k(s_i, s_j)$. Then the Markov chain with memory has the formal presentation $f_{i_1 j_1 k_1} f_{i_2 j_2 k_3} \dots f_{i_n j_n k_n} \dots$, where $j_l = k_{l-1}$, $i_l = j_{l-1}$ for all $l = 2, 3, \dots$. Accepting the last sequence as the element $\mathcal{F}_{i_1 i_2 \dots}$, $i_k = 1, 2, \dots, m$, $k = 1, 2, \dots$, of the space $\mathcal{F}$ with the metric $\delta(\mathcal{F}_{i_1 i_2 \dots}, \mathcal{F}_{l_1 l_2 \dots}) = \Sigma_{k=1}^{\infty} d(s_{i_k}, s_{l_k})/2^k$ we obtain the basis common with that for the chain without memory.

Finally, consider the Markov process with the memory of the length equal to arbitrary natural number n. Then we formalize the discussion with the elements $f_{i^1, \dots, i^n}$, which consist of successive elementary events $s_{i^1}, \dots, s_{i^n}$, such that the sequence $f_{i_1^1, \dots, i_1^n} \dots, f_{i_l^1, \dots, i_l^n} \dots$ with $i_l^j = i_{l-1}^{j+1}$, $j = 1, \dots, n-1$, $l = 2, 3, \dots$, is the formalization of the chain. We obtain the structure for the dynamics research, if we accept the last sequence as an element $\mathcal{F}_{i_1^1 i_2^1 \dots}$, $i_k^1 = 1, 2, \dots, m$, $k = 1, 2, \dots$, of the space $\mathcal{F}$, placing stress on the events with the indices i_k^1, $k = 1, 2, \dots$.

It is clear that for all cases, regardless of whether they are with memory or without memory, we have constructed one and the same space $\mathcal{F}$ of elements $\mathcal{F}_{i_1 i_2 \dots}$, $i_k = 1, 2, \dots, m$, $k = 1, 2, \dots$, with the distance $\delta(\mathcal{F}_{i_1 i_2 \dots}, \mathcal{F}_{l_1 l_2 \dots}) = \Sigma_{k=1}^{\infty} d(s_{i_k}, s_{l_k})/2^k$. This is why the space is the main object of analysis for the presence of chaos in the random process in what follows.

Complete the dynamics with the special mapping $\varphi: \mathcal{F} \to \mathcal{F}$ such that

$$\varphi(\mathcal{F}_{i_1 i_2 \dots i_n \dots}) = \mathcal{F}_{i_2 i_3 \dots i_n \dots}, \tag{5.13}$$

for each element of the set. The map φ is in the paradigm of the Bernoulli shift [52, 58], known for the symbolic dynamics. It is said to be the *abstract similarity map* [8] as it is convenient to describe fractals which are determined through the abstract self-similarity. The abstract similarity map surpasses the Bernoulli shift in many senses. This is detailed in chapter 1.

From the definitions it implies that the appearance of values of the map is the same as for the members of the state space, which are ordered successively. Consequently, if one proves that the map is one of the chaotic types in the literature, then we have to recognize that the chaos is proper for the stochastic dynamics if appropriate probability arguments are provided.

The following sets,

$$\mathcal{F}_{i_1 i_2 \dots i_n} = \bigcup_{j_k = 1,2,\dots,m} \mathcal{F}_{i_1 i_2 \dots i_n j_1 j_2 \dots}, \tag{5.14}$$

where indices $i_1, i_2, \dots, i_n$, are fixed, were introduced in [7, 8].

It is clear that

$$\mathcal{F} \supseteq \mathcal{F}_{i_1} \supseteq \mathcal{F}_{i_1 i_2} \supseteq \dots \supseteq \mathcal{F}_{i_1 i_2 \dots i_n} \supseteq \mathcal{F}_{i_1 i_2 \dots i_n i_{n+1} \dots}, \; i_k = 1, 2, \dots, m, \; k = 1, 2, \dots,$$

that is, the sets form a nested sequence.

Since one can verify that

$$\varphi^n(\mathcal{F}_{i_1 i_2 \dots i_n}) = \mathcal{F}, \tag{5.15}$$

for arbitrary natural number n and $i_k = 1, 2, \dots, m, \; k = 1, 2, \dots$, there is a reason to call φ a *similarity map*.

We will say that for the sets $\mathcal{F}_{i_1 i_2 \dots i_n}$ the *diameter condition* is valid, if

$$\max_{i_k = 1,2,\dots,m} \operatorname{diam} \mathcal{F}_{i_1 i_2 \dots i_n} \to 0 \text{ as } n \to \infty, \tag{5.16}$$

where $\operatorname{diam}(A) = \sup\{\delta(\mathbf{x}, \mathbf{y}) \colon \mathbf{x}, \mathbf{y} \in A\}$ for a set A in $\mathcal{F}$.

Define the function $\delta(A, B) = \inf\{\delta(\mathbf{x}, \mathbf{y}) \colon \mathbf{x} \in A, \mathbf{y} \in B\}$ for two nonempty sets A and B in $\mathcal{F}$. Set $\mathcal{F}$ satisfies the *separation condition* of degree n if there exist a positive number ε_0 and a natural number n such that for arbitrary indices $i_1 i_2 \dots i_n$ one can find indices $j_1 j_2 \dots j_n$ such that

$$\delta\left(\mathcal{F}_{i_1 i_2 \dots i_n}, \mathcal{F}_{j_1 j_2 \dots j_n}\right) \geqslant \varepsilon_0. \tag{5.17}$$

Next we will formulate two theorems on Poincaré and Devaney chaos. Verification of the assertions is given in [8].

Theorem 5.5.1. [7, 8] *If the diameter and separation conditions are valid, then the dynamics* $(\mathcal{F}, \delta, \phi)$ *is chaotic in the sense of Devaney.*

The last theorem claims that the similarity map φ possesses the three ingredients of Devaney chaos, namely the density of periodic points, transitivity and sensitivity. A point $\mathcal{F}_{i_1 i_2 i_3 \dots} \in \mathcal{F}$ is periodic with period n if its index consists of endless repetitions of a block of n terms.

In [10, 11] Poisson stable motion is utilized to distinguish chaotic behavior from periodic motions in Devaney and Li–Yorke types. The dynamics is given the named Poincaré chaos. The next theorem shows that the Poincaré chaos is valid for the similarity mapping. The domain-structured chaos unites models which admit all types of theoretical chaos such as Poincaré, Li–Yorke and Devaney [7, 8], but the first one is most convenient and dominant to confirm that chaos is proper for the random dynamics.

Let us give the definition of the unpredictability in the sense of the dynamics $(\mathcal{F}, \delta)$.

Definition 5.5.1. [10] *A point $\mathcal{F}_{i_1 i_2 \ldots}$ from the set $\mathcal{F}$ is unpredictable, if there exist a positive number ε_0 and sequences t_m, s_m, of natural numbers both of which diverge to infinity such that $\varphi^{t_m}(\mathcal{F}_{i_1 i_2 \ldots i_k \ldots}) \to \mathcal{F}_{i_1 i_2 \ldots i_k \ldots}$ as $m \to \infty$ and $\delta(\varphi^{t_m + s_m}(\mathcal{F}_{i_1 i_2 \ldots i_k \ldots}), \varphi^{t_m}(\mathcal{F}_{i_1 i_2 \ldots i_k \ldots})) \geqslant \varepsilon_0$ for each natural number m.*

Theorem 5.5.2. [7, 8] *If the diameter and separation conditions are valid, then the similarity map possesses Poincaré chaos.*

Proof. Let us fix an unpredictable point of the dynamics as a member $\mathcal{F}_{i_1 i_2 \ldots i_k \ldots}$ of $\mathcal{F}$, which index sequence $i_1 i_2 \ldots i_k \ldots$ consists of subsequently chosen strings all of length one, then all possible strings of length two, etc, to infinity. The sample of this construction is seen for the two symbolic alphabet cases in the following diagram,

$$(\underbrace{0\ 1}_{1\text{ block}}\ |\underbrace{00\ 01\ 10\ 11}_{2\text{ blocks}}|\underbrace{000\ 001\ 010\ 011\ \ldots}_{3\text{ blocks}}|\ldots).$$

It is easily seen from the diameter condition that the point is Poisson stable. That is, there exists a sequence of integers, t_m, divergent to infinity such that $\varphi^{t_m}(\mathcal{F}_{i_1 i_2 \ldots i_k \ldots}) \to \mathcal{F}_{i_1 i_2 \ldots i_k \ldots}$ as $n \to \infty$. Consider the property of unpredictability. We will show that there is a sequence s_m divergent to infinity such that $\delta(\varphi^{t_m + s_m}(\mathcal{F}_{i_1 i_2 \ldots i_k \ldots}), \varphi^{t_m}(\mathcal{F}_{i_1 i_2 \ldots i_k \ldots})) \geqslant \varepsilon_0$, where ε_0 is the separation constant for the space $(\mathcal{F}, \delta)$. Assume in contrast that there is no such sequence. Then $\delta(\varphi^{t_m + i}(\mathcal{F}_{i_1 i_2 \ldots i_k \ldots}), \varphi^{t_m}(\mathcal{F}_{i_1 i_2 \ldots i_k \ldots})) < \varepsilon_0$, for all natural i. This contradicts the diameter condition and the choice of the initial point, since it means the absence of the initial string of length n in the index sequence of the point $\varphi^{t_m}(\mathcal{F}_{i_1 i_2 \ldots i_k \ldots})$. □

It is easily seen that for a finite metric space (S, d) the definition 5.5.1 has the next form, which is the most convenient for the present research.

Definition 5.5.2. *A bounded sequence $s_i \in S$, $i = 0, 1, \ldots$, is called unpredictable if there exist a positive number ε_0 and sequences t_n, s_n, $n = 1, 2, \ldots$, of positive integers both of which diverge to infinity such that $s_{i+t_n} = s_i$ for each bounded interval of integers, if n is sufficiently large, and $d(s_{t_n + s_n}, s_{t_n}) \geqslant \varepsilon_0$ for each natural number n.*

Let us fix an unpredictable realization of the chain, which can be determined as follows. Consider an unpredictable point $\mathcal{F}^*_{i_1 i_2 \ldots}$ of the map φ. Fix the sequence, $s^* = \{s^*_{i_k}\}_k$, which is the corresponding unpredictable realization of the Markov chain. Due to the definition 5.5.2 and theorem 5.5.2, the following assertion is valid.

Theorem 5.5.3. *Each finite realization of the Markov chain coincides with an arc of* s^*. *That is, the unpredictable realization occurs in each experiment of the chain and is a certain event.*

Thus one has to recognize that each experiment of the Markov chain, with and without memory, produces an arc of an unpredictable orbit. This result can be considered as the main achievement of this research. It agrees with the principle result of the ergodic theory [57] that a single trajectory proves the behavior of the whole dynamics and all other trajectories. In fact, we can say that it is a fixed unpredictable orbit. In other words, this is reproduction of the chaos as a certain event. This is what guarantees the irregularity of each finite sample path of the chain.

Now let us consider the random dynamics for deterministic sensitivity. For the discussion let us fix an infinite sequence of elements of the state space. The probability that a simulated realization will coincide with the sequence forever is equal to zero. Consequently, the complementary event that an experiment with the same start state will diverge from the sequence at some finite time is equal to one. Due to the metric it means that the sensitivity is present for the random dynamics with probability one.

To finalize the relationship of the stochastic dynamics and deterministic chaos, we conclude that there are elements of the deterministic chaos which are certain events such as the appearance of an unpredictable sequence as a finite realization and there are ingredients of chaos such as the sensitivity, which are most probable, but not deterministic, i.e. not certain events. Possibly, it is useful to say that chaos is proper for the random processes with probability one altogether.

5.5.3 An example: a random walk

Consider the following Markov chain as an example. Let the real-valued scalar dynamics $X_{n+1} = X_n + Y_n$, $n \geqslant 0$, be given such that $Y_n = \{-1, 1\}$ is a random variable. Since we expect the chaotic dynamics realizations to be bounded, the special chain with boundaries is constructed. That is, the probability distribution $P(1) = P(-1) = 1/2$, if $X_n \neq 1, 4$, and certain events $Y_n = -1$, if $X_n = 4$, and $Y_n = 1$, if $X_n = 1$. To satisfy the construction of the current research, we will make the following agreements. First of all, denote $s_0 = 1$, $s_1 = 2$, $s_2 = 3$, $s_3 = 4$. Consider the state space $S = \{s_1, s_2\}$. Introduce the following events, $f_{11} = \{s_1, s_0, s_1\}$, $f_{12} = \{s_1, s_2\}$, $f_{21} = \{s_2, s_1\}$, $f_{22} = \{s_2, s_3, s_2\}$. It is clear that $p_{i1} + p_{i2} = 1$, $i = 1, 2$, and all the probabilities are equal to one half. That is, despite the differences to the theoretical chain construction in the main body of the chapter, we have the sequences $f_{i_1 j_1} f_{i_2 j_2} \ldots f_{i_n j_n} \ldots$, with $i_k = j_{k-1}$ for all $k = 2, 3, \ldots$, and we are in the circumstances of domain-structured chaos. Consequently, there is unpredictability. To visualize an unpredictable realization we will draw the graph of the function $\varphi(t) = X_n$, if $t \in [n/10, (n+1)/10)$, $0 \leqslant t \leqslant 60$. According to theorem 5.6.4 it is an arc of an unpredictable sequence. The graph of the function is seen in figure 5.4. It illustrates an unpredictable sequence, the sample path of the random walk.

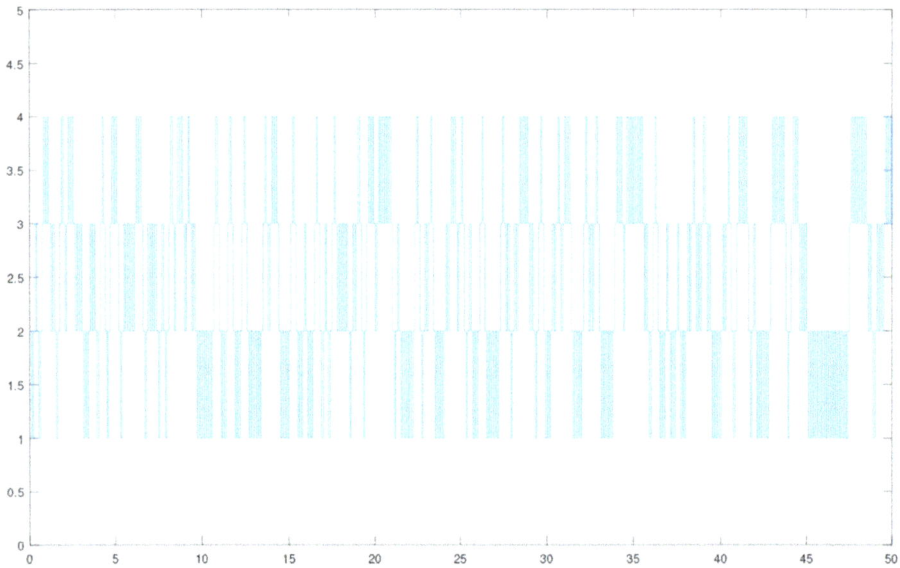

Figure 5.4. The graph of the function $\varphi(t)$ as a visualization of the unpredictable realization of the chain X_n, $n \geqslant 0$. The vertical lines are drawn for better visibility.

5.5.4 Conclusions

The outcome of the research is the existence of an unpredictable sequence as a realization of the Markov chain, and the sequence appears as a finite realization of each experiment of the process. That is, the appearance of the sequence is a certain event for the stochastic dynamics. At the same time the sensitivity does not appear deterministically, but with probability one. From this point of view one can say that the deterministic chaos is an event for the stochastic dynamics with probability one. This result is true for many other discrete-time random processes, for example the Bernoulli scheme. The significant use of the investigation is that one can unite methods of deterministic chaos with those for stochastic dynamics. Many other opportunities may appear. Among the methods are the control and synchronization of chaos [29]. We have discussed that the sensitivity is present with probability one. Evidently it is true for the Bernoulli scheme [14] and other dynamics, which can be proved for the abstract similarity dynamics. Next our study will relate Markov processes with continuous time, as well as unbounded. This also is connected to many other random processes. Our results shed more light on the Markov chains as ergodic processes, since we have shown that there is the uncountable set of realizations, which are unpredictable orbits of the abstract similarity dynamics and each of them is dense in the set of all realizations. [10].

5.6 Modular chaos in random processes

We apply the method of domain-structured chaos for a dynamics with orbits whirling among infinitely many modules. Thus the conservative conditions of the existence of chaos are weakened to model irregular dynamics. The research proves that Poincaré chaos is of exceptional use for the analysis of stochastic processes. Examples illustrating the results are provided.

5.6.1 Introduction

The paradigm of chaos relies on two distance antagonists, proximality and separation. They utilize metrical or topological closeness and divergence in the description considering either couples of orbits [23, 38] or a single motion [10]. In the current research, we have renounced the idea that transitivity and density of periodic motions of a class of motions have to be described metrically and even topologically. It is suggested, for the first time in the literature, that the closeness can be considered not in the metrical sense, but through *indexing*, although separation still depends entirely on the distance. That is, the antagonists still are present, but not in the way of conventional theories. The process of indexing is adjacent to the similarity map. More precisely, the indexing is the instrument for how the mapping is to be chaotic. Trajectories move through different sets and even various metric spaces when the time increases and each of the sets admits its own chaotic structure. This is why we say that the *modular chaos* is the focus of the chapter or, more precisely, *modular domain-structured chaos*.

The method of domain-structured chaos was initiated in [2, 7, 8]. Results were obtained which depend on dynamics in a single metrical or topological space, fractals and bounded random processes such as Bernoulli schemes [2, 8, 13]. Here the method is enriched in two senses. First, we consider infinite multiples of chaotically structured modules. Second, the conservative chaos conditions are weakened by considering indexing for closeness. This helps us to determine unbounded deterministic chaotic dynamics and find chaotic structures for continuous-time random processes, bounded as well as unbounded. We have accepted the presence of chaos in discrete-time stochastic dynamics if realizations coincide with orbits of the modular similarity map and explain its presence for continuous-time random processes. One must point out that the concepts of Poincaré chaos and unpredictable sequences [10, 11] are central to the exploration.

We hope that the approach will be developed further by considering multidimensional indexing and infinitely dimensional dynamics. Additionally, the definitions can be considered for processes which are stochastic only partially, that is for hybrid dynamics, where intervals of deterministic and/or stochastic behavior are combined.

Let us shed more light on the probabilistic essence of the chaos and the random process relations. Considering the modular Poincaré chaos, one can see that there exists a modular unpredictable realization such that any finite realization is an arc of the unpredictable one. Consequently, we have to conclude that in any experiment we observe the modular unpredictable orbit as a certain event. This is the main argument for why we compare the two types of dynamics in our research.

At the basis of our investigations are the effective methods which have been developed for chaos investigation: *abstract similarity dynamics* and *abstract similarity chaos*.

The research on similarity goes back to Leibniz, who introduced the notions of recursive self-similarity [59]. The idea of a self-similar set was first considered by

Moran [42]. He gave a mathematical definition of a geometric construction as a collection of sets satisfying specific conditions. Further, definitions of self-similarity and related problems were discussed in the papers [17, 25, 27, 31–33, 37, 43, 50, 55, 56]. Our research is based on a point-set structure in metric spaces and this is why we call it *abstract self-similarity* [8]. It does not depend on any special functions and the *similarity map* is naturally combined with the concept. We suppose that next extension of the current results can be obtained on the basis of papers [17, 25, 27, 31, 32, 37, 43, 55].

Potential applications of the modular chaos naturally relate to problems of control and synchronization [29]. The theory of stochastic differential equations [45] and Markov chains [28] can be enriched, if modular chaos investigation results are involved in the study by utilizing methods of replication of chaos and unpredictable dynamics [4, 9, 10, 13].

It is of strong interest to compare deterministic and random dynamics. More precisely, to find features of chaos in stochastic processes as well as to recognize probabilistic features in deterministic motions. If the first one is considered, for example in papers [47, 48], in the paper [26] arguments of statistics are provided to show deterministic features in random processes. Apparently, our research is in the second stream, where the random processes can be assigned with the properties of chaos. Our results depend not on the statistical parameters, but on the concept which has been introduced and developed during the last four years—Poincaré chaos [10, 11].

5.6.2 Modular domain-structured chaos

Let a countable collection of compact metric spaces (F_j, d_j), $j = 1, 2, \dots$, be given with distances d_j. Assume that for every set F_j the following presentation exists

$$\mathcal{F}^j = \left\{F^j_{i_1 i_2 \dots i_n \dots}: i_k = 1, 2, \dots, m, \ k = 1, 2, \dots\right\}, j = 1, 2, \dots, \tag{5.18}$$

where m is a natural number not smaller than two and common for all these sets. It means that each element of the set F_j is labeled through at least one member of $\mathcal{F}^j$ and every element of set $\mathcal{F}^j$ presents a member from F_j. The uniqueness is not necessarily required for the relation. That is, if $\mathcal{F}^j_{i_1 i_2 \dots i_n \dots}$ and $\mathcal{F}^j_{j_1 j_2 \dots jn, \dots}$ present the same element f from F_j it is not necessary that $i_n = j_n$ for all $n = 1, 2, \dots$. Moreover, we determine functions $\delta_j(\mathcal{F}^j_{i_1 i_2 \dots i_n \dots}, \mathcal{F}^j_{j_1 j_2 \dots jn, \dots}) = d_j(f_1, f_2)$, $j = 1, 2, \dots$, if $\mathcal{F}^j_{i_1 i_2 \dots i_n \dots}$ and $\mathcal{F}^j_{j_1 j_2 \dots jn, \dots}$ correspond to elements f_1 and f_2 of the set F_j, respectively, such that $\delta_j(\mathcal{F}^j_{i_1 i_2 \dots i_n \dots}, \mathcal{F}^j_{j_1 j_2 \dots jn, \dots}) = 0$ for different presentations of the same point in F_j.

The following sets are required:

$$\mathcal{F}^j_{i_1 i_2 \dots i_n} = \bigcup_{j_k = 1, 2, \dots, m} \mathcal{F}^j_{i_1 i_2 \dots i_n j_1 j_2 \dots}, \tag{5.19}$$

where indices $i_1, i_2, \dots, i_n$, are fixed. It is clear that

$$\mathcal{F}^j \supseteq \mathcal{F}^j_{i_1} \supseteq \mathcal{F}^j_{i_1 i_2} \supseteq \dots \supseteq \mathcal{F}^j_{i_1 i_2 \dots i_n} \supseteq \mathcal{F}^j_{i_1 i_2 \dots i_n i_{n+1}} \dots, \ i_k = 1, 2, \dots, m, \ k = 1, 2, \dots.$$

That is, the sets form a nested sequence.

We will say that the *diameter condition* is valid for sets $\mathcal{F}^j_{i_1 i_2 \ldots i_n}$, $j = 1, \ldots$, if

$$\max_{i_1 i_2 \ldots i_n} \operatorname{diam} \mathcal{F}^j_{i_1 i_2 \ldots i_n} \to 0 \text{ as } n \to \infty, \tag{5.20}$$

where $\operatorname{diam}(A) = \sup\{\delta_j(\mathbf{x}, \mathbf{y}): \mathbf{x}, \mathbf{y} \in A\}$, for a set A in $\mathcal{F}_j$.

Determine the function $\delta_j(A, B) = \inf\{\delta_j(\mathbf{x}, \mathbf{y}): \mathbf{x} \in A, \mathbf{y} \in B\}$ for two nonempty bounded sets A and B in $\mathcal{F}_j$. The set $\mathcal{F}^j$, $j = 1, 2, \ldots$, satisfies the *separation condition* of degree $n = n(j)$, if there exist a positive number ε_0^j such that for arbitrary indices $i_1 i_2 \ldots i_n$ one can find indices $j_1 j_2 \ldots j_n$, which satisfy

$$\delta_j\left(\mathcal{F}^j_{i_1 i_2 \ldots i_n}, \mathcal{F}^j_{j_1 j_2 \ldots j_n}\right) \geqslant \varepsilon_0^j. \tag{5.21}$$

If the diameter and separation conditions are valid, then $\mathcal{F}^j$ is said to be a *abstract chaotic set* for F_j, $j = 1, 2, \ldots$, and a *chaotic module* for F.

We shall say that the union, $\mathcal{F}$, of all modules $\mathcal{F}_j$, $j = 1, 2, \ldots$, is a *modular chaotic structure* for the union F of all sets F_j, $j = 1, 2, \ldots$, if the diameter condition is valid for all $j = 1, 2, \ldots$, and there exists a positive separation constant $\varepsilon_0 = \inf_{j=1,2,\ldots} \varepsilon_0^j$.

Let us introduce the maps φ_j: $\mathcal{F}^j \to \mathcal{F}^{j+1}$, $j = 1, 2, \ldots$, such that

$$\varphi_j\left(\mathcal{F}^j_{i_1 i_2 \ldots i_n \ldots}\right) = \mathcal{F}^{j+1}_{i_2 i_3 \ldots i_n \ldots}. \tag{5.22}$$

In what follows we refer to the map φ on the set $\mathcal{F}$, which is specified by equations (5.22). That is, $\varphi(\mathcal{F}^j) = \varphi_j(\mathcal{F}^j)$, $j = 1, 2, \ldots$. Moreover, accept that $\delta(\mathcal{F}^j_{i_1 i_2 \ldots i_n \ldots}, \mathcal{F}^j_{j_1 j_2 \ldots j_n \ldots}) = \delta_j(\mathcal{F}^j_{i_1 i_2 \ldots i_n \ldots}, \mathcal{F}^j_{j_1 j_2 \ldots j_n \ldots})$, $j = 1, 2, \ldots$.

Considering iterations of the map, one can verify that

$$(\varphi_{j+n}(\ldots(\varphi_j\left(\mathcal{F}^j_{i_1 i_2 \ldots i_n}\right))\ldots) = \varphi^n\left(\mathcal{F}^j_{i_1 i_2 \ldots i_n}\right) = \mathcal{F}^{j+n+1}, \tag{5.23}$$

for arbitrary natural number n and $i_k = 1, 2, \ldots, m$, $k = 1, 2, \ldots$. The relations (5.22) and (5.23) give us a reason to call φ an *affine similarity map*, since the shifts for the lower and upper indices are in common. One can also call the map a *modular similarity map*. Moreover, the number n is called the *order of similarity*. It is clear that the map cannot be considered as a modification of the Bernoulli map, since it is not a shift map. There is no direction in the phase space and shifts occur in the indices, but not in elements of the dynamics. Analogously to the research in [8], we will call the set $\mathcal{F}$ a *modular abstract similarity set*.

The set $\mathcal{F}^j$ for a fixed $j = 1, 2, \ldots$ is an *abstract similarity set* [8]. If $\mathcal{F}^j$ does not depend on j the modular similarity map φ is known as the *abstract similarity map* [2, 7].

Let us determine the map Φ: $F_j \to F_{j+1}$, $j = 1, 2, \ldots$, a geometrical representation of the similarity map. Fix a number j and a point f from F_j. Then $\Phi(f)$ is the collection of all elements in F_{j+1}, the labels of which are values of the function φ at labels of f. It is clear that Φ is a set-valued mapping, but we shall call it a map for simplicity. In what follows we shall describe chaotic properties of the map Φ in terms of the dynamics of the map φ.

We shall say that an element $\mathcal{F}^j_{i_1i_2\ldots i_n\ldots}$ is in the (k, m)-neighborhood of a point $\mathcal{F}^p_{s_1s_2\ldots s_n\ldots}, j < p$, if $\mathcal{F}^{j+k}_{i_{k+1}i_{k+2}\ldots i_{k+m}} = F^p_{s_1s_2\ldots s_m}$. In other words, $j + k = p, i_{k+1} = s_1, \ldots, i_{k+m} = s_m$.

Since the concepts are similar to that of the Bernoulli shift [58], known for the symbolic dynamics, it is easy to see that the following chaotic properties are valid.

A point $\mathcal{F}^j_{i_1i_2i_3\ldots}, j = 1, 2, \ldots$, is periodic with period p if its lower index consists of endless repetitions of a block of p terms. The map φ is said to be *modular-periodic*, since $\varphi^{kp}(\mathcal{F}^j_{i_1i_2\ldots i_n\ldots}) = \mathcal{F}^{j+kp}_{i_1i_2\ldots i_n\ldots}$ for any natural k if the point $\mathcal{F}^j_{i_1i_2\ldots i_n\ldots}$ is p-periodic.

Similarly, the map is *modular dense* or *modular-transitive*, since there exists the lower index $i_1i_2i_3\ldots$, such that for any sequence $j_1j_2j_3\ldots$ and natural j and l one can find positive integers m and $k, j < k$, such that $\varphi^m(\mathcal{F}^j_{i_1i_2\ldots i_n\ldots}) = \mathcal{F}^k_{j_1j_2\ldots j_n\ldots}$, and $i_s = j_s$, $s = 1, \ldots, l$. In other words $\mathcal{F}^j_{i_1i_2\ldots i_n\ldots}$ is in the (m, l)-neighborhood of $\mathcal{F}^k_{j_1j_2\ldots j_n\ldots}$.

Finally, one can say that the periodic points are *modular dense* in $\mathcal{F}$, since for any point $\mathcal{F}^k_{j_1j_2\ldots j_n\ldots}$ of the set and natural numbers l and $m \leqslant k$ one can find a periodic point $\mathcal{F}^m_{i_1i_2\ldots i_n\ldots}$ such that it is in the $(k - m, l)$-neighborhood of $\mathcal{F}^k_{j_1j_2\ldots j_n\ldots}$.

It is clear that couples $(\mathcal{F}^j, \delta_j), j = 1, 2, \ldots$, are not, in general, metric spaces. However, the map φ is convenient to extend all attributes of Poincaré, Li–Yorke and Devaney chaos [13, 23, 38] for the dynamics, since the spaces (F_j, d_j) are metric. We shall say that the map φ is *modular chaotic in the Devaney sense* if it is dense with respect to modular-periodic trajectories, modular-transitive and *modular-sensitive*. The modular-sensitivity means that there exists a positive number ε_0 such that for each element $\mathcal{F}^j_{i_1i_2\ldots i_n\ldots}$ and arbitrary positive κ one can find an element $\mathcal{F}^j_{j_1j_2\ldots j_n\ldots}$ and a natural number k, which satisfy $\delta(\varphi^k(\mathcal{F}^j_{i_1i_2\ldots i_n\ldots}), \varphi^k(\mathcal{F}^j_{j_1j_2\ldots j_n\ldots})) > \varepsilon_0$, despite $\delta(\mathcal{F}^j_{i_1i_2\ldots i_n\ldots}, \mathcal{F}^j_{j_1j_2\ldots j_n\ldots}) < \kappa$. Finally, the map Φ is *modular chaotic in the sense of Devaney*, if the corresponding map φ is modular chaotic in the same sense.

Theorem 5.6.1. *If $\mathcal{F}$ is a modular chaotic structure then the dynamics of Φ is modular chaotic in the sense of Devaney.*

Proof. Modular-transitivity and density for modular-periodicity are valid, due to the definitions. For modular-sensitivity fix a point $\mathcal{F}^j_{i_1i_2\ldots} \in \mathcal{F}$ and a positive number κ. Due to the diameter and separation conditions there exist an integer k and element $\mathcal{F}^j_{i_1i_2\ldots i_k j_{k+1}j_{k+2}\ldots}$ such that $0 < \delta(\mathcal{F}^j_{i_1i_2\ldots i_ki_{k+1}\ldots}, \mathcal{F}^j_{i_1i_2\ldots i_kj_{k+1}\ldots}) < \kappa$ and $\delta(\mathcal{F}^j_{i_{k+1}i_{k+2}\ldots i_{k+n}}, \mathcal{F}^j_{j_{k+1}j_{k+2}\ldots j_{k+n}}) > \varepsilon_0$. This proves the sensitivity. □

In [10, 11] Poisson stable motion is utilized to distinguish chaotic behavior from periodic motions in Devaney and Li–Yorke type chaos. The dynamics is called Poincaré chaos. Let us provide the definition adapted for the dynamics of the current research. A point $\mathcal{F}^j_{i_1i_2\ldots}$ from $\mathcal{F}, j = 1, 2, \ldots$, is said to be the *modular unpredictable point* for the map φ if there exist a positive number ε_0 and two unbounded sequences of natural numbers, κ_n and ζ_n, $n = 1, 2, \ldots$, such that for arbitrary natural number l one can find sufficiently large κ_n such that $\varphi^{\kappa_n}(\mathcal{F}^j_{i_1 i_2 \ldots}) = \mathcal{F}^{j+\kappa_n}_{j_1 j_2 \ldots}$, where $j_s = i_s$, $s = 1, \ldots, l$, and $j_{\kappa_n+\zeta_n} \neq i_{\kappa_n+\zeta_n}$.

Following the definition of Poincaré chaos [10] we shall say that the dynamics of the map φ is *modular chaotic in the sense of Poincaré* if there is a modular unpredictable point for each $j = 1, 2, \ldots$. Moreover, we call the unpredictable point itself an unpredictable orbit of the dynamics. It is natural to accept that the map Φ is *modular chaotic in the sense of Poincaré* if the corresponding map φ is modular chaotic in the sense of Poincaré.

Theorem 5.6.2. *If $\mathcal{F}$ is a modular chaotic structure then the map Φ possesses modular chaos in the sense of Poincaré.*

The proof of the last theorem is based on the technique of theorem 5.6.1 and similar to that of the verification of theorem 3.1.1 applied to the modular similarity map.

In addition to the modular Devaney and Poincaré chaos, the chaos of Li–Yorke type [38] can be defined. In addition to the modular-periodic points of the dynamics, assume that there exists an uncountable *scrambled set* of non-periodic points in $\mathcal{F}^j$, $j = 1, 2, \ldots$, such that

$$(\alpha) \limsup_{k\to\infty} \delta\Big(\varphi^k\big(\mathcal{F}^j_{i_1 i_2 \ldots}\big), \varphi^k\big(\mathcal{F}^j_{i_j j_2 \ldots}\big)\Big) > 0,$$

and

$$(\beta) \liminf_{k\to\infty} \delta\Big(\varphi^k\big(\mathcal{F}^j_{i_1 i_2 \ldots}\big), \varphi^k\big(\mathcal{F}^j_{i_j j_2 \ldots}\big)\Big) = 0,$$

for each couple in the set, if $\delta(\mathcal{F}^j_{i_1 i_2 \ldots}, \mathcal{F}^j_{i_j j_2 \ldots}) \neq 0$. Finally, assume that for each point $\mathcal{F}^j_{i_1 i_2 \ldots}$ of the set and a periodic point $\mathcal{F}^j_{i_j j_2 \ldots}$ it is true that $\limsup_{k\to\infty}\delta (\varphi^k(\mathcal{F}^j_{i_1 i_2 \ldots}), \varphi^k(\mathcal{F}^j_{i_j j_2 \ldots})) > 0$. The conditions ($\alpha$) and ($\beta$) are called *modular-proximality* and *modular-frequent separation* properties for the *modular Li–Yorke chaos*, respectively.

We will say that for the sets $\mathcal{F}^j_{i_1 i_2 \ldots i_n}$, $j = 1, \ldots$, the *strong diameter condition* is valid, if

$$\sup_j \max_{i_1 i_2 \ldots i_n} \operatorname{diam} \mathcal{F}^j_{i_1 i_2 \ldots i_n} \to 0 \text{ as } n \to \infty. \tag{5.24}$$

We shall say that the union, $\mathcal{F}$, of all modules $\mathcal{F}_j$, $j = 1, 2, \ldots$, is a *strong modular chaotic structure* for the union F of all sets F_j, $j = 1, 2, \ldots$, if the strong diameter condition is valid for all $j = 1, 2, \ldots$, and there exists a positive separation constant $\varepsilon_0 = \inf_{j=1,2,\ldots} \varepsilon_0^j$. We shall call the map Φ *modular chaotic in the sense of Li–Yorke* if the corresponding map φ is modular chaotic in the sense of Li–Yorke.

Theorem 5.6.3. *If $\mathcal{F}$ is a strong modular chaotic structure then the map Φ is modular chaotic in the Li–Yorke sense.*

The proof of the assertion is similar to that of theorem 6.35 in [22] for the shift map defined in the space of symbolic sequences.

One can easily see that, in the case when (F_j, d_j) does not depend on j, we obtain domain-structured chaos, which was introduced in [2]. That is, modular chaos can be considered as a generalization of domain-structured chaos.

The analysis of ingredients for the three types of chaos implies that the most weak and, consequently, most flexible properties are required for Poincaré modular chaos, since their distance evaluations are not required. Nevertheless, we are confident that this must not be considered as a deficiency. The description can still provide circumstances sufficient for irregularity, which can be useful for future applications.

Example 5.6.1. *Introduce the finite metric space (F, d) of two elements f and g. Assign to f and g infinitely many labels $\mathcal{F}_{1i_2...}$ and $\mathcal{F}_{2i_2...}$, respectively, such that $i_k = 1, 2$ for all $k = 1, 2, \ldots$. To satisfy the conditions of the theoretical discussion, we consider all the modules to be identical such that $F^j = F, j = 1, 2, \ldots$. Thus a domain-structured chaos* [2] *is under discussion. Then one can see that the similarity map φ produces the chaotic dynamics of the map Φ on the finite set. To demonstrate how our suggestions can be utilized, consider the Bernoulli process with the state space (F, d). Theorem* 5.6.4 *of the next subsection or the results in* [2] *imply that each realization of the process is an arc of an unpredictable sequence.*

Example 5.6.2. *(modular chaos) Take into account metric spaces $(F^j, d_j), j = 1, 2, \ldots,$ of sequences $f = (i_1^j, i_2^j, \ldots)$, which coordinates are elements of the finite sets $(a_1^j, a_2^j, \ldots, a_m^j)$, m is a natural number, and metrics*

$$d_j(f_1, f_2) = \sum_{l=1}^{\infty} \frac{\rho_j(i_l^j, k_l^j)}{2^{l-1}},$$

where $f_1 = (i_1^j, i_2^j, \ldots)$ and $f_2 = (k_1^j, k_2^j, \ldots)$ are members of $F^j, j = 1, 2, \ldots,$ and ρ_j are distances for elements of the alphabets. Considering modules $\mathcal{F}^j, j = 1, 2, \ldots,$ as sets of labels $\mathcal{F}^j_{i_1i_2...} = (i_1^j, i_2^j, \ldots)$, we deal with the modular chaos.

To show how the domain-structured chaos is more convenient in chaos construction than the traditional method of topological conjugacy, let us consider the following example.

Example 5.6.3. *It is known that there exists the Cantor set, $\mathcal{C}$, in the section* [0, 1], *such that to each point of the set the label $\tau_{i_1i_2...}$, $i_k = 0, 1$, $k = 1, 2, \ldots,$ is assigned*

[23, 58]. *Consider the structure $\mathcal{T}$, which consists of the labels, and introduce the distance*

$$d(\tau_{i_1i_2\ldots}, \tau_{j_1j_2\ldots}) = |c_1 - c_2|,$$

where $|\cdot|$ is the absolute value and $\tau_{i_1i_2\ldots}$ with $\tau_{j_1j_2\ldots}$ are labels for points c_1 and c_2 in the set. The labeling procedure confirms that the structure is chaotic. Thus we have obtained the chaos for the geometrical object, without topological equivalence of the logistic map and the symbolic dynamics [23]. It is important that it coincides with the dynamics of the logistic map.

Let us consider another version of the discussion. One can apply the distance

$$d(\tau_{i_1i_2\ldots}, \tau_{j_1j_2\ldots}) = \sum_{k=1}^{\infty} \frac{|i_k - j_k|}{2^{k-1}}$$

for the structure. Then repeat the analysis with the distance $d(c_1, c_2) = d(\tau_{i_1i_2\ldots}, \tau_{j_1j_2\ldots})$ in $\mathcal{C}$ and the additional remark on the equivalence of the two metrics,

$$1/3|c_1 - c_2| \leqslant \sum_{k=1}^{\infty} \frac{|i_k - j_k|}{2^{k-1}} \leqslant 9/2|c_1 - c_2|,$$

to make a conclusion of the domain-structured chaos on $\mathcal{C}$.

If the last example is not completely new for chaos research, the next model has not been considered in the sense at all. We will show how a domain-structured chaos can be developed into a modular one. The results are useful for various types of hybrid systems [3, 9], which involve either moments of discontinuity or switching moments as orbits of the model.

Example 5.6.4. *(modular chaos) Consider shifts $\mathcal{C}^j$, $j = 1, 2, \ldots$, of the Cantor set $\mathcal{C}$ from the last example along the real axis by natural numbers. That is, $\mathcal{C}^j = \{c + (j - 1) | c \in \mathcal{C}\}$, $j = 1, 2, \ldots$. Accordingly, we label points of the Cantor sets as $\tau^j_{i_1i_2\ldots}$, $i_k = 0, 1, k, j = 1, 2, \ldots$. Determine the distance*

$$d_j(c_1, c_2) = d_j\left(\tau^j_{i_1i_2\ldots}, \tau^j_{j_1j_2\ldots}\right) = \sum_{k=1}^{\infty} \frac{|i_k - j_k|}{2^{k-1}}, j = 1, 2, \ldots,$$

if $\tau^j_{i_1i_2\ldots}$ and $\tau^j_{j_1j_2\ldots}$ are labels of the points c_1 and c_2 of the set $\mathcal{C}^j$, $j = 1, 2, \ldots$. By considering the similarity map

$$\varphi\left(\tau^j_{i_1i_2\ldots}\right) = \tau^{j+1}_{i_2i_3\ldots}$$

we arrange the modular chaos on the union of the Cantor sets $\mathcal{C}^j$, corresponding to modules $\mathcal{T}^j$, $j = 1, 2, \ldots$.

To find a concrete meaning for the last construction, let us consider the non-autonomous scalar equation $x_{i+1} = (i + 1) + m(x_i - i)(1 - x_i + i)$, $m > 4$. Fix $\mathcal{C}$ as *the invariant set for the map $mx(1 - x)$ in the unit section. It is clear that the union of sets $\mathcal{C}^j$, $j = 1, 2, \ldots$, is the domain for the modular chaos of the equation.*

5.6.3 Chaotic random processes

Let us describe the type of stochastic processes [24] that are the focus of this section. Consider a random process with continuous time or discrete time and with continuous or discrete state space as a family of random variables $\mathbf{X}(t)$ indexed by the parameter t with the range $\mathcal{I}$, which is either an interval of the real line or an infinite set of integers bounded from below. Realizations of the dynamics are not necessarily continuous if the time is continuous and they must not be bounded functions for both sorts of time. Denote by $S = \bigcup_{t\in\mathcal{I}} S^t$ the state space of the process and consider it with a distance d. We assume positive probability for all members of the state space and for each fix t one of the members of the state space must necessarily occur.

Suppose that for each fixed $t \in \mathcal{I}$ there exists a presentation

$$\mathcal{S}^t = \left\{ \mathcal{S}^t_{i_1 i_2 \ldots i_n \ldots} \colon i_k = 1, 2, \ldots, m, \ \ k = 1, 2, \ldots \right\}, \tag{5.25}$$

where m is a fixed natural number, common for all $t \in \mathcal{I}$, larger than or equal to two, such that each element of the set S^t is labeled as a member of the set $\mathcal{S}^t$. The indexing is in the sense of the last section and corresponding to the distance d function δ is common for all $t \in \mathcal{I}$.

Consider all possible infinite strictly increasing sequences $t_\alpha = \{t_i^\alpha\}$, $i = 1, 2, \ldots$, in $\mathcal{I}$, with $\mathcal{A}$ the set of all indexes α.

We shall say that the random process $\mathbf{X}(t)$ is *(strongly) modular chaotic* if $\mathcal{S}^\alpha = \bigcup_i \mathcal{S}^{t_i^\alpha}$ is a (strong) modular chaotic structure for $S^\alpha = \bigcup_i S^{t_i^\alpha}$ with each $\alpha \in \mathcal{A}$ and there exists a number ε_0, which is not larger than all separation constants ε_0^α, $\alpha \in \mathcal{A}$.

The choice of modular chaos for the random dynamics description is proved, since the set of all realizations for dynamics $\mathbf{X}(t_\alpha)$ with fixed $\alpha \in \mathcal{A}$ coincides with the set of all trajectories of the affine similarity map φ on the set $\mathcal{S}^\alpha$, and the dynamics of the affine map is similar to the symbolic dynamics. To argue this, let us turn the discussion to the analysis of the lower indices. Consider the space $\Sigma_m = \{(i_1 i_2 \ldots) \,|\, i_k = 1, 2, \ldots, m\}$ of infinite sequences on a finite number of symbols with the metric

$$d(s, t) = \sum_{k=1}^{\infty} \frac{|i_k - j_k|}{2^{k-1}},$$

where $(i_1 i_2 \ldots)$ and $(j_1 j_2 \ldots)$ are elements of Σ_m. The Bernoulli shift $\sigma \colon \Sigma_m \to \Sigma_m$ is defined as $\sigma(s_0 s_1 s_2 \ldots) = (s_1 s_2 s_3 \ldots)$. The map σ is continuous and (Σ_r, d) is a compact metric space [23, 58]. According to the result in [10], the symbolic dynamics admits an unpredictable point, i^*, an element of the set [11]. It is important for us that Σ_m is a closure for the unpredictable orbit. The peculiarity of the metric implies that each arc of any sequence in the space coincides with some arc of the unpredictable sequence with a shift precision. In other words, due to the finite number of indices involved in the construction, one has complete coincidence with a part of the unpredictable orbit. Moreover, the coincidence does not last forever due to the unpredictability. This is why each iteration of the Bernoulli shift σ results in what is done by the Bernoulli trial and we accept the random dynamics as modular chaotic. Consequently, the following assertion is valid for a fixed unpredictable realization of the random process.

Theorem 5.6.4. *Each finite realization of the dynamics $X(t_\alpha)$ coincides with an arc of the modular unpredictable realization. That is, the modular unpredictable realization is a certain event.*

Let us give more arguments for the presence of chaos on the basis of our research, considering the sensitivity of the dynamics. Fix an infinite sequence. Consider it as a realization. The probability that the realization will occur as the prescribed sequence is equal to zero. Hence, the complementary event, that there is a finite moment of time such that the realization changes its value from the fixed sequence value, admits the probability one. It is easy to see that it is the probability of the sensitivity. More precisely, if we start two experiments with identical states, then there exists a moment, with probability one, such that they diverge. Thus we can say that the sensitivity presents in the random dynamics with probability one. One can call the property a *random sensitivity*.

The final conclusion of the discussion is that each unpredictable realization of the chain appears at each experiment of the chain as a certain event, and the sensitivity of the dynamics occurs with probability one. Thus we cannot claim that the chaos appears as a deterministic phenomenon in the random process, but what we can say are two things. The first is that the chaos occurs with probability one and the second is that one can look for ingredients of the deterministic chaos, which are present in the dynamics as certain events. Possibly, the problem of seeking the ingredients can be beneficial for applications and looks attractive for theoretical issues.

The closure of the unpredictable realization is said to be the quasi-minimal set [53]. It contains an uncountable set of unpredictable realizations. One can say that the result of the current research is the proof that the union of all infinite realizations of the random process is a quasi-minimal set, which is a certain event. This is another argument for why we call the process chaotic.

It is clear that for the discrete-time random process it is sufficient to consider only the sequence related to the set of indices $\mathcal{I}$ itself. We do not exclude that the method of modular chaos may be of use even if the modular chaotic structure exists not for all α from $\mathcal{A}$, but for some of them.

On the basis of the last discussion one can say that stochastic processes are more 'irregular' than deterministic chaos. However, we have strong confidence that by developing the current results one can make the difference between the two concepts smaller, if not diminish entirely.

Next we will provide three examples of random processes which are modular chaotic.

Example 5.6.5. *Consider several scalar real-valued functions* $f_j(t), j = 1, \ldots, m,$ *defined on a time interval,* $\mathcal{I}$*, of real numbers. Assume that the ranges of the functions are disjoint and the minimal distance between elements of distinct ranges is larger than a positive number* ε_0*. Define the random process* $\boldsymbol{X}(t)$ *such that for each fixed* $t \in \mathcal{I}$ *it admits the value of one of the functions* f_j *with positive probability* $p_j, j = 1, \ldots, m.$ *The probabilities do not depend on t and the sum of all probabilities is equal to the unit if t is fixed.*

We have that the state space is $S = \bigcup_{t \in \mathcal{I}} S^t$*, where* $S^t = \{ f_j(t) | j = 1, \ldots, m\}$*. Let us construct the presentation*

$$\mathcal{S}^t = \left\{\mathcal{S}^t_{i_1 i_2 \ldots i_n \ldots}\colon i_k = 1, 2, \ldots, m, \;\; k = 1, 2, \ldots\right\}, \tag{5.26}$$

where $\mathcal{S}^t_{i_1 i_2 \ldots i_n \ldots} = f_{i_1}(t), \; i_j = 1, 2, \ldots, m, \;\; j = 2, \ldots.$ *One can check that the union,* $\mathcal{S}$*, of sets* $\mathcal{S}^t$*,* $t \in \mathcal{I}$*, is a modular chaotic structure for the set S and consequently the random process is modular chaotic.*

Example 5.6.6. *(numerical simulation) Let us consider the random process* $\boldsymbol{X}(t)$*, which is equal to t or* $-t$ *for each fixed* $t \in [1, 4]$*, with probability* 1/2*. It is clear that the dynamics is a continuous-time random process with continuous state space. Fix the parameter value t and obtain that* $S^t = \{t, -t\}$ *consists of two elements, and the distance between the two points is equal* $|t - (-t)| = 2t \geqslant \varepsilon_0 = 2$*. Let us construct a chaotic structure for the set* S^t*. Denote* $S^t_{i_1 i_2 \ldots} = t$ *with* $i_1 = 1$*, and* i_j *is equal to* 1 *or* 2 *for all* $j \geqslant 2$*. Similarly,* $S^t_{i_1 i_2 \ldots} = -t$ *with* $i_1 = 2$ *and* i_j *is equal to* 1 *or* 2 *for all* $j \geqslant 2$*. According to discussion in the last example the random process is modular chaotic. To illustrate the dynamics we consider the sequence* $t_i = i/100, \; i = 100, 2, \ldots, 400,$ *randomly determine values* $\boldsymbol{X}(t_i)$ *and draw the graph of the piece-wise constant function* φ *equal to* $\boldsymbol{X}(t_i)$ *on the interval* $[i/100, (i+1)/100), \; i = 100, \ldots, 399.$ *The graph is seen in figure* 5.5 *and it approximates one of the realizations of the random dynamics.*

Example 5.6.7. *(discrete-time dynamics) Let us determine functions* $f_i(t) = t$ *and* $g_i(t) = -t$ *on the interval* $[i/10, (i+1)/10)$ *for each* $i = 10, 11, \ldots.$ *Construct the metric space of continuous functions,* $S^i = \{ f_i(t), g_i(t)\}, \; t \in [i/10, (i+1)/10),$ *for each* $i = 10, 11, \ldots,$ *with the distance* $d(f_i(t), g_i(t)) = \max_t |t - (-t)|$*. Introduce the discrete-time random process* $\boldsymbol{X}(i), \; i = 10, 11, \ldots,$ *which is equal to the function* $f_i(t)$ *or* $g_i(t)$ *with probability* 1/2 *on the interval* $[i/10, (i+1)/10)$*. One can easily see that the random dynamics is modular chaotic, since one can construct an appropriate modular chaotic structure. The graph of a realization,* ψ*, on the interval* [1, 4] *is seen in figure* 5.6*. For better visibility of the dynamics the vertical lines connecting pieces of the graph are drawn.*

5.6.4 Notes

We have pursued three goals in the current research. The first is to reveal signs of deterministic chaos in random phenomena and the second to maximize the presence. New definitions of chaos to fulfill the two tasks have been developed. All the three

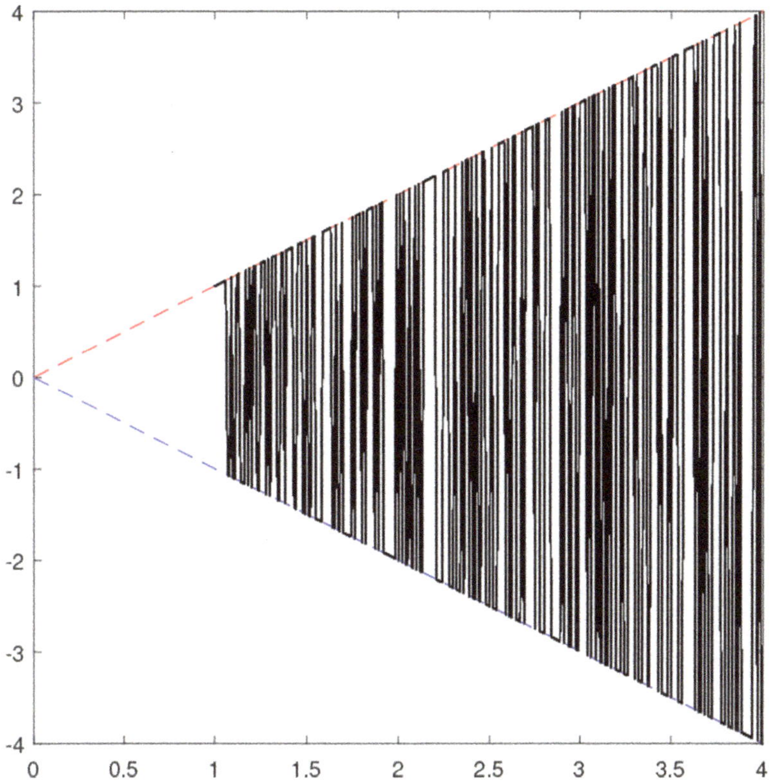

Figure 5.5. The graph of the function φ in interval $[1, 4]$. The red dashed line is the graph of the identity function t and the blue dashed line is the graph of function $-t$.

are considered in this research for random processes with discrete and continuous time. The application of the modular chaos concept to random dynamics analysis is sensible, since the set of all realizations of a random dynamics coincides with the set of all trajectories of the chaos. To prove the phenomenon the notion of domain-structured chaos, which has been considered in our previous papers, is extended. The new type of chaos has been developed such that the notions of transitivity and density of periodic orbits are free of metrical or topological features and are based on the comprehension that two points are close if they belong to the same set by indexing. Moreover, trajectories move among different sets when the time increases and each of the sets admits its own chaotic structure. It is important that the concept helped us to find a chaotic structure for continuous-time random processes, bounded as well as unbounded. We accept the phenomenon as a sign of deterministic chaos in the dynamics, since the realizations of the corresponding discrete-time random process are the same as the trajectories of the deterministic modular similarity map introduced in the current research. The concept can be considered for processes which are stochastic only partially in time, that is hybrid dynamics, through combinations of intervals where a dynamics is deterministic or stochastic.

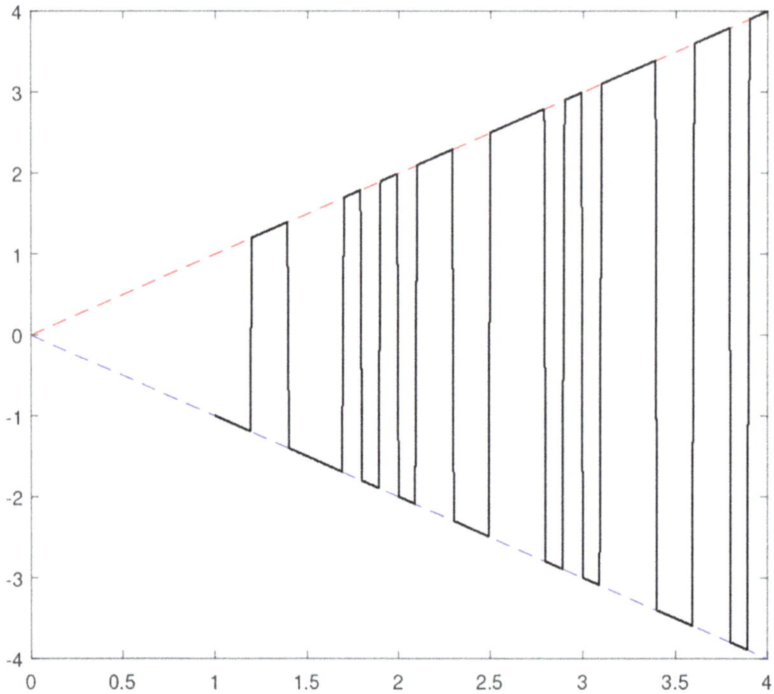

Figure 5.6. The graph of the function ψ in interval [1, 4]. The red dashed line is the graph of the identity function t and the blue dashed line is the graph of function $-t$. The vertical lines connecting the pieces of the graph are drawn for better visibility.

We hope that all this may generate new approaches inside the stochastic dynamics theory [45] as well as for non-stationary processes and for Markov chains. Our research results may help in the development of chaos in non-equilibrium processes [44]. The methods of chaotic dynamics, e.g. synchronization and control [29], can be extended to dynamics with probability. Moreover, one can expect that the results for symbolic dynamics considered in the course of the present research can give new effects for entropy [1, 59], harmonic analysis, discrete mathematics, probability and operator algebras [33].

References

[1] Adler R, Konheim A and McAndrew M 1965 Topological entropy *Trans. Am. Math. Soc.* **114** 309–19

[2] Akhmet M 1912 Domain-structured chaos in discrete random processes arXiv: 1912.10478 [math.DS]

[3] Akhmet M 2020 *Almost Periodicity, Chaos and Asymptotic Equivalence* (Berlin: Springer)

[4] Akhmet M 2020 Domain structured chaos in Markov chains arXiv: 2010.14625v1 [math.DS]

[5] Akhmet M 2021 Abstract hyperbolic chaos *Discontinuity Nonlinearity Complex* (in press)

[6] Akhmet M 2021 Modular chaos for random processes *Discontinuity Nonlinearity Complex* (in press)

[7] Akhmet M and Alejaily E M 2019 Domain-structured chaos in a Hopfield neural network *Int. J. Bifurcat. Chaos* **29** 1950205
[8] Akhmet M and Alejaily E M 2021 Abstract similarity, chaos and fractals *Discrete Contin. Dyn. Syst. Ser.* B **26** 2479–97
[9] Akhmet M and Fen M O 2016 *Replication of Chaos in Neural Networks, Economics and Physics* (Berlin: Springer)
[10] Akhmet M and Fen M O 2016 Poincaré chaos and unpredictable functions *Commun. Nonlinear Sci. Numer. Simul.* **48** 85–94
[11] Akhmet M and Fen M O 2016 Unpredictable points and chaos *Commun. Nonlinear Sci. Numer. Simul.* **40** 1–5
[12] Akhmet M and Fen M O 2018 Non-autonomous equations with unpredictable solutions *Commun. Nonlinear Sci. Numer. Simul.* **59** 657–70
[13] Akhmet M, Fen M O and Alejaily E M 2020 *Dynamics with Chaos and Fractals* (Berlin: Springer)
[14] Akhmet M and Tola A 2020 Unpredictable strings *Kazakh Math. J.* **20** 16–22
[15] Alekseev V M and Yakobson M V 1981 Symbolic dynamics and hyperbolic dynamic systems *Phys. Rep.* **75** 290–325
[16] Arnold L 1998 *Random Dynamical Systems* (Berlin: Springer)
[17] Bandt C and Graf S 1992 Self-similar sets 7. A characterization of self-similar fractals with positive Hausdorff measure *Proc. Am. Math. Soc.* **114** 995–1001
[18] Bowen R 1970 Markov partitions and minimal sets for axiom A diffeomorphisms *Am. J. Math.* **92** 907–18
[19] Bowen R 1970 Markov partitions for axiom A diffeomorphisms *Am. J. Math.* **92** 725–47
[20] Bunimovich L and Sinai Y 1980 Markov partitions for dispersed billiards *Commun. Math. Phys.* **78** 247–80
[21] Castaneda L B, Arunachalam V and Dharmarajs S 2012 *Introduction to Probability and Stochastic Processes with Applications* (Hoboken, NJ: Wiley)
[22] Chen G and Huang Y 2011 *Chaotic Maps: Dynamics, Fractals and Rapid Fluctuations, Synthesis Lectures on Mathematics and Statistics* (San Rafael, CA: Morgan and Claypool)
[23] Devaney R L 1989 *An Introduction to Chaotic Dynamical Systems* (Canada; USA: Addison-Wesley)
[24] Doob J L 1953 *Stochastic Processes* (York: Wiley)
[25] Edgar G A 1990 *Measure, Topology, and Fractal Geometry* (New York: Springer)
[26] Eisencraft M, Monteiro L H A and Soriano D C 2017 White Gaussian chaos *IEEE Commun. Lett.* **21** 1719–22
[27] Falconer K J 1995 Sub-self-similar sets *Trans. Am. Math. Soc.* **347** 3121–9
[28] Gagniuc P A 2017 *Markov Chains: From Theory to Implementation and Experimentation* (Hoboken, NJ: Wiley)
[29] Gonzáles-Miranda J M 2004 *Synchronization and Control of Chaos* (London: Imperial College Press)
[30] Hajek B 2015 *Random Processes for Engineers* (Cambridge, UK: Cambridge University Press)
[31] Hata M 1985 On the structure of self-similar sets *Japan J. Appl. Math.* **2** 381–414
[32] Hutchinson J 1981 Fractals and self-similarity *Indiana Univ. Math. J.* **30** 713–47
[33] Jorgensen P E T 2006 *Analysis and Probability: Wavelets, Signals, Fractals Graduate Texts in Mathematics* vol 234 (New York: Springer)

[34] Karlin S and Taylor H E 1975 *A First Course in Stochastic Processes* (New York: Academic)
[35] Klenke A 2014 *Probability Theory* (London: Springer)
[36] Lasota A and Mackey M C 1985 *Probabilistic Properties of Deterministic Systems* (Cambridge: Cambridge University Press)
[37] Lau K S, Ngai S M and Rao H 2001 Iterated function systems with overlaps and the self-similar measures *J. London Math. Soc.* **63** 99–116
[38] Li T Y and Yorke J A 1975 Period three implies chaos *Am. Math. Mon.* **82** 985–92
[39] Markov A A 1971 *Extension of the Limit Theorems of Probability Theory to a Sum of Variables Connected in a Chain* vol 1 (New York: Wiley)
[40] Masoller C, Schifino A C and Sicardi R L 1995 Characterization of strange attractors of Lorenz model of general circulation of the atmosphere *Chaos Soliton. Fract.* **6** 357–66
[41] Meyn S and Tweedie R L 2009 *Markov Chains and Stochastic Stability* (Cambridge: Cambridge University Press)
[42] Moran P A P 1946 Additive functions of intervals and Hausdorff measure *Proc. Camb. Philos. Soc.* **42** 15–23
[43] Ngai S M and Wang Y 2001 Hausdorff dimension of overlapping self-similar sets *J. London Math. Soc.* **63** 665–72
[44] Nicolis G and Prigogine I 1989 *Exploring Complexity: An Introduction* (New York: Freeman)
[45] Oksendal B K 2003 *Stochastic Differential Equations: An Introduction with Applications* (Berlin: Springer)
[46] Ornstein D 1970 Bernoulli shifts with the same entropy are isomorphic *Adv. Math.* **4** 337–52
[47] Ornstein D 1995 In what sense can a deterministic system be random? *Chaos Soliton. Fract.* **5** 139–41
[48] Ornstein D 2004 Kolmogorov, random processes, and Newtonian dynamics *Russ. Math. Surv.* **59** 313–7
[49] Ornstein D and Weiss B 1998 On the Bernoulli nature of systems with some hyperbolic structure *Ergodic Theory Dyn. Syst.* **18** 441–56
[50] Pesin Y and Weiss H 1996 On the dimension of deterministic and random Cantor-like sets, symbolic dynamics, and the Eckmann–Ruelle conjecture *Commun. Math. Phys.* **182** 105–53
[51] Ponce G and Varao R 2019 *An Introduction to the Kolmogorov–Bernoulli Equivalence* (Cham: Springer)
[52] Robinson C 2012 *An Introduction to Dynamical Systems: Continuous and Discrete* (Providence, RI: American Mathematical Society)
[53] Sell G R 1971 *Topological Dynamics and Ordinary Differential Equations* (London: Van Nostrand Reinhold)
[54] Shields P 1973 *The Theory of Bernoulli Shifts* (Chicago, IL: University of Chicago Press)
[55] Spear D W 1992 Measure and self-similarity *Adv. Math.* **91** 143–57
[56] Stella S 1992 On Hausdorff dimension of recurrent net fractals *Proc. Am. Math. Soc.* **116** 389–400
[57] Walters P 1982 *An Introduction to Ergodic Theory* (New York: Springer)
[58] Wiggins S 1988 *Global Bifurcation and Chaos, Analytical Methods* (New York: Springer)
[59] Zmeskal O, Dzik P and Vesely M 2013 Entropy of fractal systems *Comput. Math. Appl.* **66** 135–46

Part IV

Non-autonomous differential equations

Chapter 6

Unpredictability in differential equations

The reason why the current and following chapter are included in this book, the main subject of which is abstract similarity chaos, is that the abstract dynamics can be realized as domain structured chaos in many models. We are particularly interested in a single unpredictable motion, since it is an isolated product of the chaotic processes with very specific and constructive properties. One of the processes is abstract similarity dynamics. The motions are specified as unpredictable functions of continuous time and unpredictable sequences as functions of discrete time. Consequently, one can/must focus on unpredictable functions as solutions of various dynamical equations as well as perturbations for those equations. In our research it is considered for deterministic models as well for models with chance.

In this chapter the existence and uniqueness of unpredictable solutions in the dynamics of quasilinear systems is investigated. Moreover, strongly unpredictable functions are considered as solutions. Simulations illustrating the chaos are provided. This research is another stream of our activity, where we attempt to bring chaos and the theory of differential equations closer and make them more amenable to each other. This line of research was initiated in [1, 8–15, 22–28], where the method of replication of chaos is developed. This method has been proved to be very powerful. It is based on definitions of chaos with continuous time, while all previous descriptions used discrete time. The results for synchronization and an extension of chaos have been obtained in the papers, which are stronger in many senses than those obtained previously in the literature.

6.1 Preliminaries

It is worth noting that a large number of results are obtained for periodic, quasi-periodic and almost periodic solutions of differential equations due to the established mathematical methods and important applications in the literature [29, 31–34]. On the other hand, recurrent and Poisson stable solutions are also crucial for the theory of differential equations [30, 36, 37]. Our proposal can revive the interest of

doi:10.1088/978-0-7503-3507-2ch6

specialists in differential equations theory, due to the verification of unpredictability, which requires a more sophisticated technique than for recurrent and Poisson stable solutions. That is, the problem of the existence of unpredictable solutions is a challenging one. Unpredictable solutions can be investigated for various types of differential equations such as partial differential equations, evolution equations and hybrid systems. Thus a new approach for the extension of chaos into many types of dynamics is suggested in our study. In [37] a general method of the comparability of functions by the character of their recurrence was suggested for differential equations. In particular, it is popular for solutions with Poisson stability. In this chapter, and in our papers [4–7], we introduce and apply the approach to prove the existence of Poisson stable solutions, which is much more straightforward and consequently effective than the approach in [30, 37]. It can be applied, in the future, to various types of dynamical equations.

A new type of oscillation, unpredictable functions, was introduced in [3]. It was shown in [2, 4, 5, 7] that the existence of an unpredictable solution implies *Poincaré chaos* for a special dynamics in a functional space. Consequently, the study of unpredictable solutions is as beneficial as the study of chaos. In the current chapter we consider oscillations with all coordinates being unpredictable. Due to the chaos, irregularity is observable in the oscillations. Another significant novelty is that perturbations are assumed to be nonlinear functions of the time and space variables, and they are strongly unpredictable in the time variable.

Throughout this chapter $\mathbb{R}$ and $\mathbb{N}$ will stand for the set of real and natural numbers, respectively. Additionally, the norm $\|u\|_1 = \sup_{t\in\mathbb{R}}\|u(t)\|$, where $\|u\| = \max_{1\leqslant i\leqslant p} |u_i|$, $u = (u_1, \dots, u_p)$, $u_i \in \mathbb{R}$, $i = 1, 2, \dots, p$, will be used. Correspondingly, for a square matrix $A = \{a_{ij}\}$, $i, j = 1, 2, \dots, p$, the norm $\|A\| = \max_{i=1,\dots,p} \sum_{j=1}^{p}|a_{ij}|$, will be utilized.

The following two definitions are starting points of our research.

Definition 6.1.1. [2] *A uniformly continuous and bounded function* v: $\mathbb{R} \to \mathbb{R}^p$ *is unpredictable if there exist positive numbers* ε_0, σ *and sequences* t_n, u_n, *both of which diverge to infinity such that* $v(t + t_n) \to v(t)$ *as* $n \to \infty$ *uniformly on compact subsets of* $\mathbb{R}$ *and* $\|v(t + t_n) - v(t)\| \geqslant \varepsilon_0$ *for each* $t \in [u_n - \sigma, u_n + \sigma]$ *and* $n \in \mathbb{N}$.

Definition 6.1.2. [19] *A continuous and bounded function* $f(t, x)$: $\mathbb{R} \times G \to \mathbb{R}^p$, $f = (f_1, f_2, \dots, f_p)$, $G \subset \mathbb{R}^p$ *is a bounded domain and is unpredictable in* t *if it is uniformly continuous in* t *and there exist positive numbers* ε_0, σ *and sequences* t_n, u_n *both of which diverge to infinity such that* $\sup_G \|f(t + t_n, x) - f(t, x)\| \to 0$ *as* $n \to \infty$ *uniformly on compact sets in* $\mathbb{R}$ *and* $\inf_G \|f(t + t_n, x) - f(t, x)\| \geqslant \varepsilon_0$ *for* $t \in [u_n - \sigma, u_n + \sigma]$ *and* $n \in \mathbb{N}$.

The current study contains two principal novelties. The first is that strongly unpredictable solutions are considered alongside unpredictable ones. Second, we consider nonlinear perturbations, which are functions that are unpredictable in the time variable. To this end we shall need the following two notions, which are analogs of the last two definitions.

Definition 6.1.3. [19] *A uniformly continuous and bounded function* $v: \mathbb{R} \to \mathbb{R}^p$, $v = (v_1, \dots, v_p)$, *is strongly unpredictable if there exist positive numbers* ε_0, σ *and sequences* t_n, u_n *both of which diverge to infinity such that* $v(t + t_n) \to v(t)$ *as* $n \to \infty$ *uniformly on compact subsets of* $\mathbb{R}$ *and* $|v_i(t + t_n) - v_i(t)| \geqslant \varepsilon_0$ *for each* $t \in [u_n - \sigma, u_n + \sigma]$, $i = 1, \dots, p$ *and* $n \in \mathbb{N}$.

Definition 6.1.4. [19] *A continuous and bounded function* $f(t, x): \mathbb{R} \times G \to \mathbb{R}^p$, $f = (f_1, f_2, \dots, f_p)$, $G \subset \mathbb{R}^p$ *is a bounded domain and is strongly unpredictable in t if it is uniformly continuous in t and there exist positive numbers* ε_0, σ *and sequences* t_n, u_n, *both of which diverge to infinity such that* $\sup_G \|f(t + t_n, x) - f(t, x)\| \to 0$ *as* $n \to \infty$ *uniformly on compact sets in* $\mathbb{R}$ *and* $\inf_{[u_n - \sigma, u_n + \sigma] \times G} |f_i(t + t_n, x) - f_i(t, x)| > \varepsilon_0$ *for all* $i = 1, \dots, p$ *and* $n \in \mathbb{N}$.

Comparing the definitions 6.1.1 and 6.1.3 as well as definitions 6.1.2 and 6.1.4 one can find that an unpredictable function may admit some of coordinates which are not unpredictable scalar valued functions, while each coordinate of a strongly unpredictable function is an unpredictable function. That is, the set of all strongly unpredictable functions is a subclass of unpredictable functions.

Next let us recall the definition of a Poisson stable function [36, p 124] adapted to our case.

Definition 6.1.5. *A continuous and bounded function* $\vartheta: \mathbb{R} \to \mathbb{R}^m$ *is positively Poisson stable if there exists a sequence* $\{t_n\}$, $t_n \to \infty$ *as* $n \to \infty$, *such that* $\vartheta(t + t_n) \to \vartheta(t)$ *as* $n \to \infty$ *uniformly on compact subsets of* $\mathbb{R}$.

By comparing definitions 6.1.1 and 6.1.5 one can see that any unpredictable function is Poisson stable. The Poisson stable function is a specification of the eponymous point considered for dynamical systems in [36, p 85] and [35, p 344, 345].

The following lemmas can be useful for applications of our results.

Lemma 6.1.1. [16] *Suppose that* $\phi(t): \mathbb{R} \to \mathbb{R}$ *is an unpredictable function. Then the function* $\phi^3(t)$ *is unpredictable.*

Proof. One can find numbers $\varepsilon_0 > 0$, $\delta > 0$ and sequences $\{t_n\}$, $\{u_n\}$ both of which diverge to infinity such that $\phi(t + t_n) \to \phi(t)$ as $n \to \infty$ uniformly on compact subsets of $\mathbb{R}$ and $\|\phi(t + t_n) - \phi(t)\| \geqslant \varepsilon_0$ for each $t \in [u_n - \delta, u_n + \delta]$ and $n \in \mathbb{N}$. It is easy to check that $\phi^3(t + t_n) \to \phi^3(t)$ as $n \to \infty$ uniformly on compact subsets of $\mathbb{R}$, since it follows from the uniform continuity of the cubic function on a compact set.

Fix a natural number n. Let us show that for $t \in [u_n - \delta, u_n + \delta]$ the inequality $\|\phi(t + t_n) - \phi(t)\| \geqslant \varepsilon_0$ implies $\|\phi^3(t + t_n) - \phi^3(t)\| \geqslant \varepsilon_0^3/4$.

We have that

$$|\phi^3(t+t_n) - \phi^3(t)| = \frac{1}{2}|\phi(t+t_n) - \phi(t)|[\phi^2(t+t_n) + \phi^2(t) + (\phi(t+t_n) + \phi(t))^2]$$
$$\geqslant \frac{1}{2}(\phi^2(t+t_n) + \phi^2(t))\varepsilon_0.$$

Consider the function $F(a, b) = a^2 + b^2$ for $|a - b| \geqslant \varepsilon_0$. The minimum of F occurs at the points (a, b) with $|a| = |b| = \varepsilon_0/2$. Therefore, $|\phi^3(t+t_n) - \phi^3(t)| \geqslant \varepsilon_0^3/4$ for $t \in [u_n - \delta, u_n + \delta]$. □

Lemma 6.1.2. [16] *If the function $\phi(t)$: $\mathbb{R} \to \mathbb{R}$ is unpredictable, then the function $\phi(t) + c$, where c is a constant, is also unpredictable.*

Proof. There exist positive numbers ε_0, δ and sequences $\{t_n\}$, $\{u_n\}$ both of which diverge to infinity such that $\|\phi(t+t_n) - \phi(t)\| \to 0$ as $n \to \infty$ uniformly on compact subsets of $\mathbb{R}$ and $\|\phi(t+t_n) - \phi(t)\| \geqslant \varepsilon_0$ for each $t \in [u_n - \delta, u_n + \delta]$ and $n \in \mathbb{N}$. Let us denote $\omega(t) = \phi(t) + c$. Then we have that $\|\omega(t+t_n) - \omega(t)\| = \|\phi(t+t_n) - \phi(t)\| \to 0$ as $n \to \infty$ uniformly on compact subsets of $\mathbb{R}$ and $\|\omega(t+t_n) - \omega(t)\| = \|\phi(t+t_n) - \phi(t)\| \geqslant \varepsilon_0$ for each $t \in [u_n - \delta, u_n + \delta]$ and $n \in \mathbb{N}$. Therefore the function $\phi(t) + c$ is unpredictable. □

The main object of the current chapter is the system of quasilinear differential equations

$$x'(t) = Ax(t) + f(t, x), \tag{6.1}$$

where $t \in \mathbb{R}$, $x \in \mathbb{R}^p$, p is a fixed natural number, all eigenvalues of the constant matrix $A \in \mathbb{R}^{p\times p}$ have negative real parts, f: $\mathbb{R} \times G \to \mathbb{R}^p, f = (f_1, \dots, f_p)$, $G = \{x \in \mathbb{R}^p, \|x\| < H\}$, where H is a positive number. It is true that there exist two real numbers $K \geqslant 1$ and $\gamma < 0$ such that $\|e^{At}\| \leqslant Ke^{\gamma t}$ for all $t \geqslant 0$.

The following conditions will be needed in this chapter:

(C1) The function $f(t, x)$ is strongly unpredictable in the sense of definition 6.1.4.

(C2) There exists a positive constant L such that the function $f(t, x)$ satisfies the inequality $\| f(t, x_1) - f(t, x_2)\| \leqslant L \| x_1 - x_2 \|$ for all $t \in \mathbb{R}$, $x_1, x_2 \in G$.

(C3) $\gamma < -\frac{KM}{H}$.

(C4) $\gamma < -KL$.

Definition 6.1.4 implies that there exists a positive number M such that $\sup\limits_{\mathbb{R}\times G} \| f(t, x)\| = M < \infty$.

Our purpose is to prove that system (6.1) possesses a unique strongly unpredictable solution, provided that the function $f(t, x)$ is strongly unpredictable in t. Moreover, we prove that the solution is uniformly globally exponentially stable. Additionally, the existence of an unpredictable solution, which is not strongly unpredictable, is considered for the system.

6.2 Strongly unpredictable solutions

Two principal features are at the basis of this section. The first is that all coordinates of the solution are unpredictable functions. That is, solutions are *strongly unpredictable*. Second, perturbations are *strongly unpredictable functions* in the time variable.

Let U be the set of all uniformly continuous functions $\psi(t) = (\psi_1, \psi_2, \dots, \psi_p)$, $\psi_i \in \mathbb{R}$, $i = 1, 2, \dots, p$, such that $\| \psi \|_1 < H$, and $\psi(t + t_n) \to \psi(t)$ as $n \to \infty$ uniformly on each closed and bounded interval of the real axis, where sequence t_n is the same as for function $f(t, x)$ in system (6.1).

According to the theory of differential equations [34], a function $\omega(t) = (\omega_1, \omega_2, \dots, \omega_p)$ bounded on the whole real axis is a solution of system (6.1) if and only if the integral equation

$$\omega(t) = \int_{-\infty}^{t} e^{A(t-s)} f(s, \omega(s))ds, \ t \in \mathbb{R}, \tag{6.2}$$

is satisfied.

Lemma 6.2.1. [19] *Suppose that conditions* $(C1) - (C4)$ *are valid, then the system* (6.1) *possesses a unique solution* $\omega(t) \in U$.

Proof. Define an operator Π on U such that

$$\Pi\psi(t) = \int_{-\infty}^{t} e^{A(t-s)} f(s, \psi(s))ds, \ t \in \mathbb{R}. \tag{6.3}$$

Fix an arbitrary function $\psi(t)$ that belongs to U. We have that

$$\|\Pi\psi(t)\| \leqslant \int_{-\infty}^{t} \|e^{A(t-s)}\| \|f(s, \psi(s))\| ds \leqslant \frac{KM}{|\gamma|},$$

for all $t \in \mathbb{R}$. Therefore, by condition (C3) it is true that $\|\Pi\psi\|_1 < H$.

Fix an arbitrary positive number ε and a closed interval $[a, b]$, $-\infty < a < b < \infty$, of the real axis. We will show that for sufficiently large n it is true that $\|\Pi\psi(t + t_n) - \Pi\psi(t)\| < \varepsilon$ on $[a, b]$. Let us choose two numbers $c < a$, and $\xi > 0$ satisfying

$$\frac{2KM}{|\gamma|} e^{\gamma(a-c)} < \frac{\varepsilon}{2}, \tag{6.4}$$

and

$$\frac{K}{|\gamma|} \xi (L + 1)[1 - e^{\gamma(b-c)}] < \frac{\varepsilon}{2}. \tag{6.5}$$

Consider n sufficiently large such that $\|f(t + t_n, x) - f(t, x)\| < \xi$ and $\|\psi(t + t_n) - \psi(t)\| < \xi$ for $t \in [c, b]$ and $x \in G$. Then the inequality

$$\|\Pi\psi(t+t_n)-\Pi\psi(t)\| \leqslant \int_{-\infty}^{c} \|e^{A(t-s)}\| \|f(s+t_n, \psi(s+t_n)) - f(s, \psi(s))\| ds$$
$$+ \int_{c}^{t} \|e^{A(t-s)}\| \|f(s+t_n, \psi(s+t_n)) - f(s, \psi(s))\| ds \leqslant \int_{-\infty}^{c} 2KMe^{\gamma(t-s)} ds \quad (6.6)$$
$$+ \int_{c}^{t} \xi(L+1)Ke^{\gamma(t-s)} ds \leqslant \frac{2}{|\gamma|} KMe^{\gamma(a-c)} + \frac{K}{|\gamma|}\xi(L+1)[1-e^{\gamma(b-c)}],$$

is correct for all $t \in [a, b]$.

By inequalities (6.4) and (6.5) it is true that $\|\Pi\psi(t+t_n)-\Pi\psi(t)\| < \varepsilon$ for $t \in [a, b]$ and, therefore, $\Pi\psi(t+t_n) \to \Pi\psi(t)$ uniformly as $n \to \infty$ on each closed and bounded interval of $\mathbb{R}$.

It is easy to verify that $\Pi\psi(t)$ is a uniformly continuous function. Thus the set U is invariant for the operator Π.

We proceed to show that the operator Π: $U \to U$ is contractive. Let $u(t)$ and $v(t)$ be members of U. Then we obtain that

$$\|\Pi u(t) - \Pi v(t)\| \leqslant \int_{-\infty}^{t} \|e^{A(t-s)}\| \|f(s, u(s)) - f(s, v(s))\| ds$$
$$\leqslant \int_{-\infty}^{t} Ke^{\gamma(t-s)} L \|u(s) - v(s)\| ds \leqslant \frac{KL}{|\gamma|} \|u(t) - v(t)\|_1,$$

for all $t \in \mathbb{R}$. Therefore, the inequality $\|\Pi u - \Pi v\|_1 \leqslant \frac{KL}{|\gamma|} \|u - v\|_1$ holds and according to condition (C4) the operator Π: $U \to U$ is contractive. The next task is to show that the space U is complete. Consider a Cauchy sequence $\phi_k(t)$ in U, which converges to a limit function $\phi(t)$ on $\mathbb{R}$. Fix a closed and bounded interval $I \subset \mathbb{R}$. We have that

$$\|\phi(t+t_p) - \phi(t)\| < \|\phi(t+t_n) - \phi_k(t+t_n)\|$$
$$+ \|\phi_k(t+t_n) - \phi_k(t)\| + \|\phi_k(t) - \phi(t)\|. \quad (6.7)$$

Now one can take sufficiently large n and k such that each term on the right-hand side of equation (6.7) is smaller than $\frac{\varepsilon}{3}$ for an arbitrary small ε and $t \in I$. The inequality implies that $\|\phi(t+t_n) - \phi(t)\| < \varepsilon$ on I. That is the sequence $\phi(t+t_n)$ uniformly converges to $\phi(t)$ on I. Likewise, one can check that the limit function is bounded and uniformly continuous [34]. The completeness of U is proved. From the contraction mapping theorem there exists the unique fixed point, $\omega(t) \in U$, of the operator Π, which is the unique bounded solution of the system (6.1). The lemma is proved. □

Theorem 6.2.1. [19] *If conditions (C1)–(C4) are fulfilled, then system* (6.1) *admits a unique uniformly globally exponentially stable strongly unpredictable solution.*

Proof. According to lemma 6.2.1, system (6.1) has a unique solution $\omega(t) \in U$. What is left is to verify that the unpredictability property is valid.

One can find a positive number κ, natural numbers l, k and $j = 1, \dots, p$, such that the following inequalities are valid:

$$\kappa < \sigma, \tag{6.8}$$

$$\|\omega(t+s) - \omega(t)\| < \frac{\varepsilon_0}{k}, \quad t \in \mathbb{R}, |s| < \kappa, \tag{6.9}$$

$$|f_j(t_n + u_n + s, x) - f_j(u_n + s, x)| \geqslant \varepsilon_0/2, \|x\| < H, |s| < \kappa, n \in \mathbb{N}, \tag{6.10}$$

$$\kappa\left(1/2 - \left(\frac{1}{l} + \frac{2}{k}\right)(L + \|A\|)\right) > \frac{3}{2l}. \tag{6.11}$$

Let the numbers κ, l, k and j as well as a number $n \in \mathbb{N}$ be fixed.

Denote $\Delta = |\omega_i(u_n + t_n) - \omega_i(u_n)|$ and consider two cases (i) $\Delta \geqslant \varepsilon_0/l$ and (ii) $\Delta < \varepsilon_0/l$ such that the remaining proof naturally falls into two parts:

(i) For the case $\Delta \geqslant \varepsilon_0/l$, fix additionally a positive number κ_1 sufficiently small for

$$\|\omega(t+s) - \omega(t)\| < \frac{\varepsilon_0}{4l}, \quad t \in \mathbb{R}, |s| < \kappa_1.$$

Therefore,

$$\begin{aligned} &\|\omega(t_n + t) - \omega(t)\| \geqslant \|\omega(t_n + u_n) - \omega(u_n)\| - \|\omega(u_n) - \omega(t)\| \\ &- \|\omega(t_n + t) - \omega(t_n + u_n)\| \geqslant \frac{\varepsilon_0}{l} - \frac{\varepsilon_0}{4l} - \frac{\varepsilon_0}{4l} = \frac{\varepsilon}{2l}, \end{aligned} \tag{6.12}$$

if $t \in [u_n - \kappa_1, u_n + \kappa_1]$ and $n \in \mathbb{N}$.

(ii) One can find that from equation (6.9) it follows that

$$\|\omega(t_n + t) - \omega(t)\| < \frac{\varepsilon_0}{l} + \frac{\varepsilon_0}{k} + \frac{\varepsilon_0}{k} = \varepsilon_0\left(\frac{1}{l} + \frac{2}{k}\right), \tag{6.13}$$

if $t \in [u_n, u_n + \kappa]$.

We obtain from equations (6.8)–(6.11) that

$$|\omega_j(t_n + t) - \omega_j(t)| \geqslant \int_{u_n}^{t} |f_j(t_n + s, \omega(t_n + s)) - f_j(s, \omega(t_n + s))| ds -$$

$$\int_{u_n}^{t} |f_j(s, \omega(t_n + s)) - f_j(s, \omega(s))| ds - \int_{u_n}^{t} |\sum_{j=1}^{p} a_{ji}[\omega_i(t_n + s) - \omega_i(s)]| ds -$$

$$|\omega_i(t_n + u_n) - \omega_i(u_n)| \geqslant \frac{\kappa}{2}\varepsilon_0 - \kappa L \varepsilon_0\left(\frac{1}{l} + \frac{2}{k}\right) - \kappa\|A\|\varepsilon_0\left(\frac{1}{l} + \frac{2}{k}\right) - \frac{\varepsilon_0}{l} \geqslant \frac{\varepsilon_0}{2l},$$

for $t \in [u_n + \kappa/2, u_n + \kappa]$.

Thus, the solution $\omega(t)$ is unpredictable.

The asymptotical stability of the solution $\omega(t)$ can be verified as stability of a bounded solution in [34]. The theorem is proved. □

6.3 Unpredictable solutions

We have considered the problem of the existence and uniqueness of strongly unpredictable solutions in the last section. In what follows we will search for quasilinear systems with unpredictable solutions, which are not strongly unpredictable. For this reason assume that the following condition is valid.

(C5) The function $f(t, x)$ is unpredictable in the sense of definition 6.1.2.

Theorem 6.3.1. [19] *Suppose that the conditions (C2)–(C5) hold. Then the system (6.1) admits a unique uniformly globally exponentially stable unpredictable solution.*

Proof. One can easily see, proceeding by following the last theorem, that there exists a unique solution $\omega(t) \in U$ for system (6.1). The solution is globally uniformly asymptotically stable. What is left is to show that the unpredictability property is valid.

We have that

$$\omega(t + t_n) - \omega(t) = \omega(u_n + t_n) - \omega(u_n) + \int_{u_n}^{t} A[\omega(s + t_n) - \omega(s)]ds + \int_{u_n}^{t} [f(t_n + s, \omega(s + t_n)) - f(s, \omega(s))]ds, \; t \in \mathbb{R}. \tag{6.14}$$

One can find a positive number κ and natural numbers l, k and $j = 1, \dots, p$, such that

$$\kappa < \sigma, \tag{6.15}$$

$$\|\omega(t + s) - \omega(t)\| < \frac{\varepsilon_0}{k}, \quad t \in \mathbb{R}, |s| < \kappa, \tag{6.16}$$

$$|f_j(t_n + u_n + s, x) - f_j(u_n + s, x)| \geqslant \varepsilon_0/2, \|x\| < H, |s| < \kappa, n \in \mathbb{N}, \tag{6.17}$$

$$\kappa\left(1/2 - \left(\frac{1}{l} + \frac{2}{k}\right)(L + \|A\|)\right) > \frac{3}{2l}. \tag{6.18}$$

Denote $\Delta = \|\omega(t_n + u_n) - \omega(u_n)\|$ and consider two alternative cases (i) $\Delta \geqslant \varepsilon_0/l$ and (ii) $\Delta < \varepsilon_0/l$.

(i) For the case $\Delta \geqslant \varepsilon_0/l$, fix additionally a positive number κ_1 sufficiently small for

$$\|\omega(t + s) - \omega(t)\| < \frac{\varepsilon_0}{4l}, \quad t \in \mathbb{R}, |s| < \kappa_1.$$

Therefore,

$$\|\omega(t_n + t) - \omega(t)\| \geqslant \|\omega(t_n + u_n) - \omega(u_n)\| - \|\omega(u_n) - \omega(t)\| - \|\omega(t_n + t) - \omega(t_n + u_n)\| \geqslant \frac{\varepsilon_0}{l} - \frac{\varepsilon_0}{4l} - \frac{\varepsilon_0}{4l} = \frac{\varepsilon}{2l}, \tag{6.19}$$

if $t \in [u_n - \kappa_1, u_n + \kappa_1]$ and $n \in \mathbb{N}$.

(ii) One can find that from equation (6.16) it follows that

$$\|\omega(t_n + t) - \omega(t)\| < \frac{\varepsilon_0}{l} + \frac{\varepsilon_0}{k} + \frac{\varepsilon_0}{k} = \varepsilon_0\left(\frac{1}{l} + \frac{2}{k}\right), \tag{6.20}$$

if $t \in [u_n, u_n + \kappa]$.

We obtain from equations (6.15)–(6.18) that

$$|\omega_j(t_n + t) - \omega_j(t)| \geqslant \int_{u_n}^{t} |f_j(t_n + s, \omega(t_n + s)) - f_j(s, \omega(t_n + s))|ds-$$

$$\int_{u_n}^{t} |f_j(s, \omega(t_n + s)) - f_j(s, \omega(s))|ds - \int_{u_n}^{t} |\sum_{j=1}^{p} a_{ji}[\omega_i(t_n + s) - \omega_i(s)]|ds-$$

$$|\omega_i(t_n + u_n) - \omega_i(u_n)| \geqslant \frac{\kappa}{2}\varepsilon_0 - \kappa L\varepsilon_0\left(\frac{1}{l} + \frac{2}{k}\right) - \kappa\|A\|\varepsilon_0\left(\frac{1}{l} + \frac{2}{k}\right) - \frac{\varepsilon_0}{l} \geqslant \frac{\varepsilon_0}{2l},$$

for $t \in [u_n + \kappa/2, u_n + \kappa]$.

Thus, the solution $\omega(t)$ is unpredictable. The theorem is proved. □

6.4 Examples

Example 6.4.1. *First we will construct an unpredictable function using the dynamics of the logistic map. Let us take into account the logistic map*

$$\lambda_{i+1} = F_\mu(\lambda_i), \tag{6.21}$$

where $i \in \mathbb{Z}$ and $F_\mu(s) = \mu s(1 - s)$. The interval $[0, 1]$ *is invariant under the iterations of equation* (7.13) *for $\mu \in (0, 4]$* [34].

It was shown in example 2.2.1 *that the logistic map* (7.13) *possesses an unpredictable solution for each $\mu \in [3 + (2/3)^{1/2}, 4]$. Let $\{\rho_i\}$, $i \in \mathbb{Z}$, be an unpredictable solution of the logistic map* (7.13) *with $\mu = 3.92$ inside the unit interval* $[0, 1]$ *and consider the function*

$$\Theta(t) = \int_{-\infty}^{t} e^{-5(t-s)/2}\Omega(s)ds, \tag{6.22}$$

where $\Omega(t)$ is a piecewise constant function defined on the real axis through the equation $\Omega(t) = \rho_i$ for $t \in [i, i + 1)$, $i \in \mathbb{Z}$. It is worth noting that $\Theta(t)$ is the unique, bounded on the whole real axis, globally exponentially stable solution of the differential equation

$$v'(t) = -\frac{5}{2}v(t) + \Omega(t).$$

Additionally, one can confirm that the function $\Theta(t)$ is such that

$$\sup_{t\in\mathbb{R}} |\Theta(t)| \leqslant \frac{2}{5},$$

and it is uniformly continuous since its derivative is bounded. Because the sequence $\{\rho_i\}$ *is unpredictable, there exist a positive number* ε_0 *and sequences* $\{\zeta_n\}$, $\{\eta_n\}$ *both of which diverge to infinity such that* $|\rho_{i+\zeta_n} - \rho_i| \to 0$ *as* $n \to \infty$ *for each i in bounded intervals of integers and* $|\rho_{\zeta_n+\eta_n} - \rho_{\eta_n}| \geqslant \varepsilon_0$ *for each* $n \in \mathbb{N}$. *Fix an arbitrary positive number* ε *and arbitrary real numbers* α, β *with* $\beta > \alpha$. *Let N be a sufficiently large natural number satisfying*

$$N \geqslant \frac{2}{5} \ln\left(\frac{6}{5\varepsilon}\right).$$

There exists a natural number n_0 *such that for each* $n \geqslant n_0$ *the inequality*

$$|\rho_{i+\zeta_n} - \rho_i| < \frac{5\varepsilon}{6}$$

holds for $i = \lfloor\alpha\rfloor - N, \lfloor\alpha\rfloor - N + 1, \ldots, \lfloor\beta\rfloor$, *where* $\lfloor\alpha\rfloor$ *and* $\lfloor\beta\rfloor$, *respectively, denote the largest integers which are not greater than* α *and* β. *Accordingly, if* $n \geqslant n_0$, *then we have*

$$|\Omega(t + \zeta_n) - \Omega(t)| < \frac{5\varepsilon}{6}, \tag{6.23}$$

for $t \in [\lfloor\alpha\rfloor - N, \lfloor\beta\rfloor + 1)$. *The last inequality implies that* $|\Theta(t + \zeta_n) - \Theta(t)| < \varepsilon$ *for* $t \in [\lfloor\alpha\rfloor, \lfloor\beta\rfloor + 1]$. *Hence,* $|\Theta(t + \zeta_n) - \Theta(t)| \to 0$ *as* $n \to \infty$ *uniformly on the interval* $[\alpha, \beta]$.

On the other hand, one can confirm for each $n \in \mathbb{N}$ *that* $|\Omega(t + \zeta_n) - \Omega(t)| \geqslant \varepsilon_0$ *for* $t \in [\eta_n, \eta_n + 1)$. *For fixed* $n \in \mathbb{N}$, *using the equation*

$$\begin{aligned}\Theta(t + \zeta_n) - \Theta(t) \;&=\; \Theta(\zeta_n + \eta_n) - \Theta(\eta_n) - \frac{5}{2}\int_{\eta_n}^{t} (\Theta(s + \zeta_n) - \Theta(s))ds \\ &\quad + \int_{\eta_n}^{t} (\Omega(s + \zeta_n) - \Omega(s))ds,\end{aligned}$$

we obtain that

$$\begin{aligned}|\Theta(\zeta_n + \eta_n + 1) - \Theta(\eta_n + 1)| \geqslant &\left|\int_{\eta_n}^{\eta_n+1} (\Omega(s + \zeta_n) - \Omega(s))ds\right| - |\Theta(\zeta_n + \eta_n) - \Theta(\eta_n)| \\ &- \frac{5}{2}\left|\int_{\eta_n}^{\eta_n+1} (\Theta(s + \zeta_n) - \Theta(s))ds\right|.\end{aligned}$$

Therefore, one can verify that

$$\sup_{t\in[\eta_n,\eta_n+1]} |\Theta(t + \zeta_n) - \Theta(t)| \geqslant \frac{2\varepsilon_0}{9}.$$

Thus, there exists a sequence $\{u_n\}$ *with* $\eta_n \leqslant u_n \leqslant \eta_n + 1$, $n \in \mathbb{N}$, *such that*

$$|\Theta(u_n + \zeta_n) - \Theta(u_n)| \geqslant \frac{2\varepsilon_0}{9}.$$

For $t \in [u_n - \delta, u_n + \delta]$, *where* $\delta = \varepsilon_0/36$, *we have*

$$|\Theta(t+\zeta_n) - \Theta(t)| \geqslant |\Theta(u_n+\zeta_n) - \Theta(u_n)| - \frac{5}{2}\left|\int_{u_n}^{t} |\Theta(s+\zeta_n) - \Theta(s)|\, ds\right| - \left|\int_{u_n}^{t} |\Omega(s+\zeta_n) - \Omega(s)|\, ds\right| \geqslant \frac{\varepsilon_0}{9}.$$

It is clear that $u_n \to \infty$ *as* $n \to \infty$. *Thus the function* $\Theta(t)$ *is unpredictable.*

Example 6.4.2. [16] *Consider the system*

$$\begin{aligned} x_1' &= -2x_1 + 2x_2 - 50\Theta(t) \\ x_2' &= x_1 - 3x_2 + 5\Theta^3(t), \end{aligned} \tag{6.24}$$

where $\Theta(t)$ *is the unpredictable function defined by equation* (6.22). *The eigenvalues of the matrix of coefficients of system* (6.24) *are* -2 *and* -0.5. *One can confirm that the perturbation function* $(-50\Theta(t), 5\Theta^3(t))$ *is unpredictable in accordance with lemma* 6.1.1. *From the main result of our paper, there is an asymptotically stable unpredictable solution* $(\varphi_1(t), \varphi_2(t))$ *of system* (6.24). *Consequently, any solution of the equation ultimately behaves irregularly. This is seen from the simulation of the solution with* $x_1(0) = 0, 18$, $x_2(0) = 0, 01$ *in figure* 6.1.

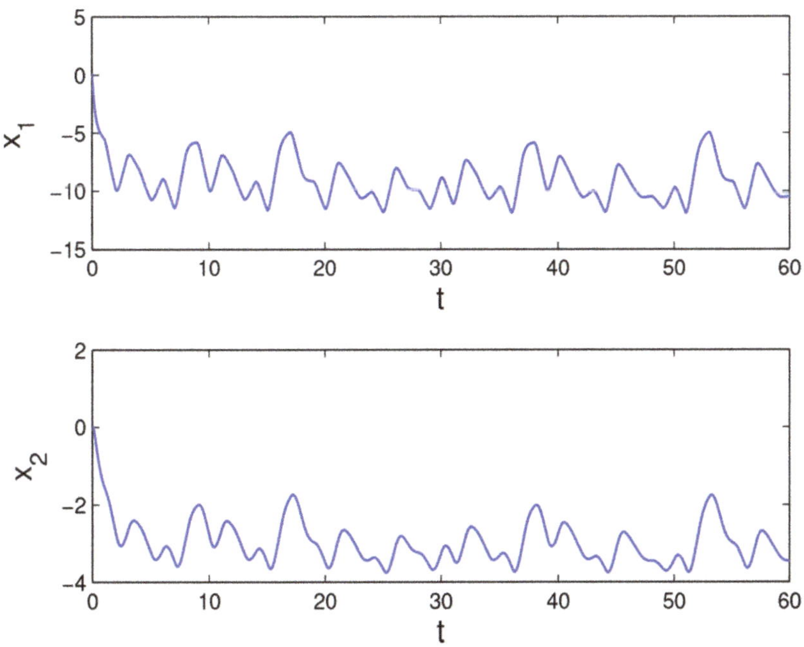

Figure 6.1. The time series of the x_1 and x_2 coordinates of system (6.24) with the initial conditions $x_1(0) = 0, 18$, $x_2(0) = 0, 01$. The figure manifests the irregular behavior of the solution.

Example 6.4.3. [19] *Consider the function* $g(t, x) = (\arctan(x) + 2)\Theta(t)$ *of two variables* t *and* x, *where* $\Theta(t)$ *is the function from example* 6.4.1. *It is easy to see that function* $g(t, x)$ *is continuously differentiable if* $t \neq i$, $i \in \mathbb{Z}$, *and bounded such that* $\sup_{\mathbb{R}\times G}|g(t, x)| = \frac{\pi}{4} + 1$. *Moreover,* $\sup_{\mathbb{R}\times G}|\frac{\partial g(t, x)}{\partial x}| = 1/2$, $t \neq i$, $i \in \mathbb{Z}$.

Let us fix an arbitrary compact interval $I \subset \mathbb{R}$ *and positive number* ε. *We have that* $|\Theta(t + t_n) - \Theta(t)| < \varepsilon$ *for* $t \in I$ *sufficiently large* n. *Consequently,*

$$|g(t + t_n, x) - g(t, x)| \leqslant |\arctan(x) + 2||\Theta(t + t_n) - \Theta(t)| < \left(\frac{\pi}{2} + 2\right)\varepsilon.$$

That is $g(t + t_n, x) \to g(t, x)$ *as* $n \to \infty$ *uniformly for* $(t, x) \in I \times G$.

On the other hand, it is true that $|\Theta(t + t_n) - \Theta(t)| \geqslant \bar{\varepsilon}_0$ *for all* $t \in [u_n - \kappa, u_n + \kappa]$ *and* $n \in \mathbb{N}$. *This is why we obtain that*

$$|g(t + t_n, x) - g(t, x)| = |\arctan(x) + 2||\Theta(t + \zeta_n) - \Theta(t)| \geqslant \left(-\frac{\pi}{2} + 2\right)\bar{\varepsilon}_0,$$

for $(t, x) \in [u_n - \kappa, u_n + \kappa] \times G$, $n \in \mathbb{N}$. *Thus* $g(t, x)$ *is (strongly) unpredictable in the t function.*

Example 6.4.4. [19] *Let us consider the system of differential equations*

$$\begin{aligned} x_1' &= -3x_1 - x_2 - x_3 + 0.51g(t, x_3) \\ x_2' &= -x_1 - 3x_2 - x_3 - 0.62g(t, x_1) \\ x_3' &= x_1 + x_2 - x_3 + 0.51g(t, x_2), \end{aligned} \tag{6.25}$$

where $g(t, x)$ *is the function from the last example. The eigenvalues of the matrix of coefficients are equal to* -2, -2 *and* -3. *One can find that conditions (C3)–(C6) are valid for system* (6.25) *with* $\gamma = -2$, $K = 6$ *and* $L = 0.31$. *According to theorem* 6.3.1 *there exists the unique asymptotically stable unpredictable solution of system* (6.25). *The simulation results for the solution of equation* (6.25) *with initial data* $\psi_1(0) = 0.05$, $\psi_2(0) = -0.1$, $\psi_3(0) = 0.15$ *are seen in figure* 6.2. *They confirm the irregularity of the dynamics.*

6.5 Notes

The current chapter is a significant part of the design for our research. This issues from the idea that differential equations are the first and main tools for the investigation of the changing world and consequently the entire world. This is why the research of chaos has to be a part of the *theory* of dynamical equations, that is, ordinary, partial, functional–differential and all other equations. In our opinion differential equations must not only be objects to test the presence of chaos, but their most general theoretic properties have to be considered as criteria for chaotic features. We have already had results of this type in the papers on the replication of chaos [1, 8–15, 22–28], where an effective input–output mechanism of chaos extension is developed for the first time in the literature. Now we involve in the mechanism a more individual actor—the unpredictable function. This chapter

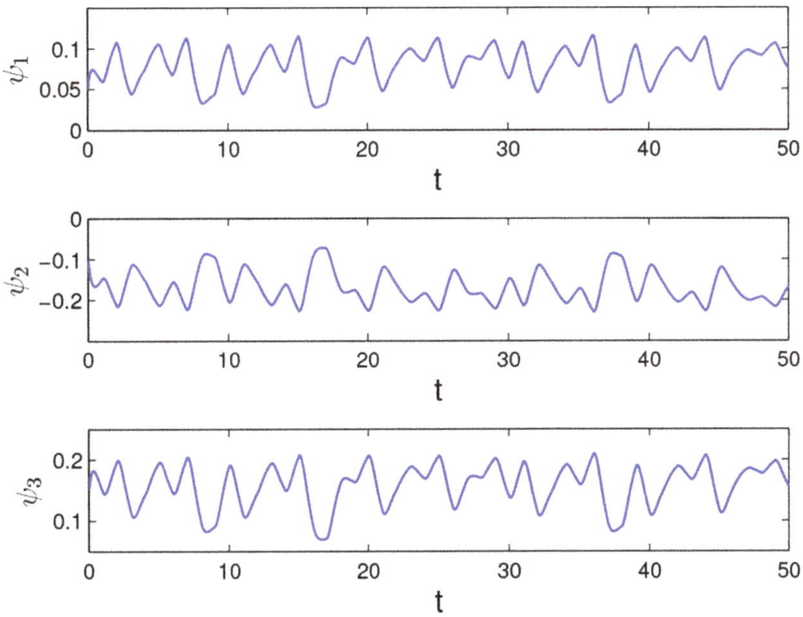

Figure 6.2. The coordinates of the solution, $\psi(t)$, of system (6.25).

contains the results of our paper [19]. The method of proof suggested in this chapter can definitely serve in various sciences. We have also realized the method for neural networks, which is presented in the next chapter and in our papers [17, 18, 20, 21].

References

[1] Akhmet M, Akhmetova Z and Fen M O 2014 Chaos in economic models with exogenous shocks *J. Econ. Behav. Organ.* **106** 95–108

[2] Akhmet M and Fen M O 2016 Poincaré chaos and unpredictable functions *Commun. Nonlinear Sci. Numer. Simul.* **48** 85–94

[3] Akhmet M and Fen M O 2016 Unpredictable points and chaos *Commun. Nonlinear Sci. Numer. Simul.* **40** 1–5

[4] Akhmet M and Fen M O 2017 Existence of unpredictable solutions and chaos *Turk. J. Math.* **41** 254–66

[5] Akhmet M and Fen M O 2018 Non-autonomous equations with unpredictable solutions *Commun. Nonlinear Sci. Numer. Simul.* **59** 657–70

[6] Akhmet M, Fen M O and Alejaily E M 2020 *Dynamics with Chaos and Fractals* (Berlin: Springer)

[7] Akhmet M, Fen M O, Tleubergenova M and Zhamanshin A 2019 Unpredictable solutions of linear differential and discrete equations *Turk. J. Math.* **43** 2377–89

[8] Akhmet M and Fen M O 2012 Chaotic period-doubling and OGY control for the forced Duffing equation *Commun. Nonlinear Sci. Numer. Simul.* **17** 1929–46

[9] Akhmet M and Fen M O 2013 Replication of chaos *Commun. Nonlinear Sci. Numer. Simul.* **18** 2626–66

[10] Akhmet M and Fen M O 2014 Chaotification of impulsive systems by perturbations *Int. J. Bifurcat. Chaos* **24** 1450078

[11] Akhmet M and Fen M O 2014 Entrainment by chaos *J. Nonlinear Sci.* **24** 411–39

[12] Akhmet M and Fen M O 2014 Generation of cyclic/toroidal chaos by Hopfield neural networks *Neurocomputing* **145** 230–9

[13] Akhmet M and Fen M O 2015 Attraction of Li–Yorke chaos by retarded SICNNs *Neurocomputing* **147** 330–42

[14] Akhmet M and Fen M O 2015 Li–Yorke chaos in hybrid systems on a time scale *Int. J. Bifurcat. Chaos* **25** 1540024

[15] Akhmet M and Fen M O 2016 Homoclinic and heteroclinic motions in economic models with exogenous shocks *Appl. Math. Nonlinear Sci.* **1** 1–10

[16] Akhmet M, Fen M O, Tleubergenova M and Zhamanshin A 2019 Poincaré chaos for a hyperbolic quasilinear system *Miskolc Math. Notes* **20** 33–44

[17] Akhmet M, Seilova R, Tleubergenova M and Zhamanshin A 2020 Shunting inhibitory cellular neural networks with strongly unpredictable oscillations *Commun. Nonlinear Sci. Numer. Simul.* **89** 105287

[18] Akhmet M, Tleubergenova M and Nugayeva Z 2020 Strongly unpredictable oscillations of Hopfield-type neural networks *Mathematics* **8** 1791

[19] Akhmet M, Tleubergenova M and Zhamanshin A 2020 Quasilinear differential equations with strongly unpredictable solutions *Carpathian J. Math.* **36** 341–9

[20] Akhmet M, Tleubergenova M and Zhamanshin A 2019 Neural networks with Poincaré chaos *ACM Int. Conf. Proc. Series* pp 34–7

[21] Akhmet M, Tleubergenova M and Zhamanshin A 2020 Inertial neural networks with unpredictable oscillations *Mathematics* **8** 1797

[22] Akhmet M U 2009 The complex dynamics of the cardiovascular system *Nonlinear Anal.* **71** e1922–31

[23] Akhmet M U 2009 Creating a chaos in a system with relay *Int. J. Qual. Theory Differ. Equ. Appl.* **3** 3–7

[24] Akhmet M U 2009 Devaney's chaos of a relay system *Commun. Nonlinear Sci. Numer. Simul.* **14** 1486–93

[25] Akhmet M U 2009 Dynamical synthesis of quasi-minimal sets *Int. J. Bifurcat. Chaos* **19** 2423–7

[26] Akhmet M U 2009 Li–Yorke chaos in the system with impacts *J. Math. Anal. Appl.* **351** 804–10

[27] Akhmet M U 2009 Shadowing and dynamical synthesis *Int. J. Bifurcat. Chaos* **19** 3339–46

[28] Akhmet M U 2010 Homoclinical structure of the chaotic attractor *Commun. Nonlinear Sci. Numer. Simul.* **15** 819–22

[29] Bohr H A 1947 *Almost Periodic Functions* (New York: Chelsea)

[30] Cheban D and Liu Z 2020 Periodic, quasi-periodic, almost periodic, almost automorphic, Birkhoff recurrent and Poisson stable solutions for stochastic differential equations *J. Differ. Equ.* **269** 3652–85

[31] Corduneanu C 2009 *Almost Periodic Oscillations and Waves* (New York: Springer)

[32] Farkas M 1994 *Periodic Motion* (New York: Springer)

[33] Fink A M 1974 *Almost Periodic Differential Equations* (Berlin: Springer)

[34] Hartman P 2002 *Ordinary Differential Equations* (New York: SIAM)

[35] Nemytskii V V and Stepanov V V 1960 *Qualitative Theory of Differential Equations* (Princeton, NJ: Princeton University Press)

[36] Sell G R 1971 *Topological Dynamics and Ordinary Differential Equations* (London: Van Nostrand Reinhold)

[37] Shcherbakov B A 1969 Poisson stable solutions of differential equations, and topological dynamics *Differ. Uravn.* **5** 2144–55 (in Russian)

Part V

Neural networks

IOP Publishing

Domain Structured Dynamics
Unpredictability, chaos, randomness, fractals, differential equations and neural networks
Marat Akhmet

Chapter 7

Unpredictable oscillations of neural networks

Sections of this chapter have been reprinted from [13], copyright (2020), with permission from Elsevier.

Oscillations are ubiquitous in brain activity and they strongly correlate with many cognitive tasks. For example, neural networks in the mammalian forebrain demonstrate several oscillatory bands covering frequencies from approximately 0.05 Hz to 500 Hz. Detailed biophysical investigations have revealed that even single neurons demonstrate the ability to resonate and oscillate, thus one can suggest that the activity within neural networks represents information [36, 50]. Oscillatory neural networks are effective for image recognition [57] and in activating network states associated with memory recall [60].

It is natural that neural oscillations became the core of the interdisciplinary research that unites neuroscience, psychophysics, biophysics, cognitive psychology, computational modeling, mathematics and physics [18]–[68]. When studying exogenous oscillations one must talk about oscillatory neural networks [31, 34, 44, 51, 57, 60, 61]. Moreover, considering the role of fluctuations in neural network activity, one must focus on the endogenous oscillations. This is why many papers are devoted to periodic and almost periodic solutions, and exponential stability for neural networks considering input/output mechanisms [2, 17, 25, 33, 37, 45–49, 65, 67, 70].

The study of chaotic oscillations in neural networks starts to be of significant interest [1]–[32]. Solutions of the chaotic systems are irregular and this is reflected by data related to experiments and observations [27, 42, 62, 64].

In [4] deterministic unpredictable functions were introduced as a new type of oscillation. The existence of unpredictable solutions proves Poincaré chaos for a Hopfield type neural network [7] and the motions admit numeric characteristics which can be useful for the analysis of neural processes. The description of such functions relies on the dynamics of unpredictable points, which were presented in [5].

The research of unpredictable solutions unites the theoretical advantages and challenges which are proper for both oscillations and chaos, and will open up many interesting prospects in neuroscience.

An unpredictable function may admit some coordinates which are not unpredictable scalar valued functions. This is why it is of great importance to consider motions which are unpredictable in *all* state dimensions, that is, *strongly unpredictable functions*.

In this chapter two new models of neural networks, Hopfield type neural networks and inertial neural networks under unpredictable perturbations, are investigated. The models are inspired by chaotic time-varying stimulus, which allows us to study the distribution of chaotic signals in neural networks by using the unpredictable input–output mechanism. Unpredictable inputs create an excitation wave of neurons that transmit chaotic signals. The first section considers the new type of oscillations for shunting inhibitory cellular neural networks. The existence and stability of strongly unpredictable solutions are investigated. In the second section, unpredictable oscillations are investigated for inertial neural networks. Unpredictable solutions that have chaotic behavior are discussed. The existence and uniqueness theorem is proved and asymptotic stability is investigated. For both models numerical simulations are presented to demonstrate the validity of the theoretical results.

7.1 SICNNs with strongly unpredictable oscillations

Shunting inhibitory cellular neural networks (SICNNs), which have been introduced by Bouzerdoum and Pinter in [21], play an exceptional role in psychophysics, robotics, adaptive pattern recognition, vision and image processing. In the last two decades many results have been published concerning the dynamics of neural networks [9, 10, 22, 24, 29, 35, 69, 74, 75].

7.1.1 The model

In its original formulation [21], the SICNN model is a two-dimensional grid of processing cells. We will follow the description in the current research. Let C_{ij} denote the cell at the (i, j) position of the lattice. Denote by $N_r(i, j)$ the r-neighborhood of C_{ij}, such that

$$N_r(i, j) = \{C_{kp}\colon \max(|k - i|, |p - j|) \leqslant r,$$
$$1 \leqslant k \leqslant m,\ 1 \leqslant l \leqslant n\},$$

where m and n are fixed natural numbers. In SICNN*s*, neighboring cells exert mutual inhibitory interactions of the shunting type. The dynamics of the cell C_{ij} is described by the following nonlinear ordinary differential equation,

$$\frac{dy_{ij}}{dt} = -a_{ij}x_{ij} - \sum_{C_{kp}\in N_r(i,j)} C_{ij}^{kp} f(x_{kp}(t))x_{ij} + v_{ij}(t), \quad (7.1)$$

where y_{ij} is the activity of the cell C_{ij}, $v_{ij}(t)$ is the external input to the cell C_{ij}, the constant a_{ij} represents the passive decay rate of the cell activity, $C_{ij}^{kp} \geqslant 0$ is the connection or coupling strength of postsynaptic activity of the cell C_{kp} transmitted to the cell C_{ij} and the activation $f(y_{kp})$ is a positive continuous function representing the output or firing rate of the cell C_{kp}, and $v_{ij}(t)$ is the external input to the cell C_{ij}. Throughout the chapter, $\mathbb{R}$ and $\mathbb{N}$ will represent the sets of real and natural numbers, respectively. Also, the norm $\|u\|_1 = \sup_{t\in\mathbb{R}}\|u(t)\|$, where $\|u\| = \max_{(i,j)} |u_{ij}|$, $u(t) = (u_{11}, \ldots, u_{1n}, \ldots, u_{m1}\ldots, u_{mn})$, $t, u_{ij} \in \mathbb{R}$, $i = 1, 2, \ldots, m$, $j = 1, 2, \ldots, n$, will be used. The following definition is an initial definition in our research.

Definition 7.1.1. [4] *A uniformly continuous and bounded function u: $\mathbb{R} \to \mathbb{R}^{m\times n}$ is unpredictable if there exist positive numbers ϵ_0, δ and sequences t_p, s_p both of which diverge to infinity such that $u(t + t_p) \to u(t)$ as $p \to \infty$ uniformly on compact subsets of $\mathbb{R}$ and $\|u(t + t_p) - u(t)\| \geqslant \epsilon_0$ for each $t \in [s_p - \delta, s_p + \delta]$ and $p \in \mathbb{N}$.*

It is seen that some coordinates of an unpredictable function may be not unpredictable scalar valued functions. This is why, in the following definition, we consider an elaborated version of the concept.

Definition 7.1.2. *A uniformly continuous and bounded function u: $\mathbb{R} \to \mathbb{R}^{m\times n}$ is strongly unpredictable if there exist positive numbers ϵ_0, δ and sequences t_p, s_p both of which diverge to infinity such that $u(t + t_p) \to u(t)$ as $p \to \infty$ uniformly on compact subsets of $\mathbb{R}$ and $|u_{ij}(t + t_p) - u_{ij}(t)| \geqslant \epsilon_0$ for all $i = 1, \ldots, m$, $j = 1, \ldots, n$, $t \in [s_p - \delta, s_p + \delta]$ and $p \in \mathbb{N}$.*

It is easy to see that any strongly unpredictable function is unpredictable in the sense of definition 7.1.1, but not vice versa. Moreover, each of the coordinates $u_{ij}(t)$, $i = 1, \ldots, m$, $j = 1, \ldots, n$, of the strongly unpredictable function is unpredictable in the sense of the definition 7.1.1.

7.1.2 The main result

Let us denote by $\mathcal{B}$ the set of functions $u(t) = (u_{11}, \ldots, u_{1n}, \ldots, u_{m1}\ldots, u_{mn})$, $t, u_{ij} \in \mathbb{R}$, $i = 1, 2, \ldots, m$, $j = 1, 2, \ldots, n$, where $m, n \in \mathbb{N}$, such that:

1. Functions $u(t)$ are uniformly continuous.
2. There exists a positive number H such that $\|u\|_1 < H$ for all $u(t) \in \mathcal{B}$.

3. There exists a sequence t_p, $t_p \to \infty$ as $p \to \infty$ such that for each $u(t)$ from $\mathcal{B}$ the sequence $u(t + t_p)$ uniformly converges to $u(t)$ on each closed and bounded interval of the real axis.

The following conditions will be needed throughout this chapter:

1. The function $v(t) = (v_{11}, \dots, v_{1n}, \dots, v_{m1}\dots, v_{mn}), t, v_{ij} \in \mathbb{R}, i = 1, 2, \dots, m, j = 1, 2, \dots, n$, in system (7.1) belongs to $\mathcal{B}$ and is strongly unpredictable such that there exist positive numbers $\delta, \epsilon_0 > 0$ and a sequence $s_p \to \infty$ as $p \to \infty$ which satisfy $|v_{ij}(t + t_p) - v_{ij}(t)| \geqslant \epsilon_0$ for all $t \in [s_p + \delta, s_p - \delta]$, $i = 1, \dots, m, j = 1, \dots, n$, and $p \in \mathbb{N}$.
2. $\gamma = \min_{(i,j)} a_{ij} > 0$.
3. There exist positive numbers m_{ij} such that $\sup_{t\in \mathrm{R}} \left|v_{ij}(t)\right| \leqslant m_{ij}$.
4. There exists a positive number m_f such that $\sup_{|s|<H} |f(s)| \leqslant m_f$.
5. There exists a positive number L such that $|f(s_1) - f(s_2)| \leqslant L\,|s_1 - s_2|$ for all $s_1, s_2, |s_1| < H, |s_2| < H$.
6. $a_{ij} > m_f \sum_{C_{kp}\in N_r(i,j)} C_{ij}^{kp}$.
7. $\dfrac{m_{ij}}{a_{ij} - m_f \sum_{C_{kp}\in N_r(i,j)} C_{ij}^{kp}} < H$.
8. $(LH + m_f)\max_{(i,j)} \sum_{C_{kp}\in N_r(i,j)} C_{ij}^{kp} < \gamma$ for all $i = 1, \dots, m, j = 1, \dots, n$.

Like the result in [26, 30], one can verify that the following assertion is valid.

Lemma 7.1.1. [13] *A function bounded on* $\mathbb{R}$ $y(t) = \left\{y_{ij}(t)\right\}$, $i = 1, \dots, m, j = 1, \dots, n$, *is a solution of SICNNs* (7.1) *if and only if the following integral equation is satisfied*

$$y_{ij}(t) = -\int_{-\infty}^{t} \mathrm{e}^{-a_{ij}(t-s)}\left[\sum_{C_{kp}\in N_r(i,j)} C_{ij}^{kp} f(y_{kp}(s))y_{ij}(s) - v_{ij}(s)\right]ds, \tag{7.2}$$

for all $i = 1, \dots, m, j = 1, \dots, n$.

Define on $\mathcal{B}$ the operator Π such that $\Pi u(t) = \left\{\Pi_{ij} u(t)\right\}$, $i = 1, \dots, m, j = 1, \dots, n$, where

$$\Pi_{ij} u(t) \equiv -\int_{-\infty}^{t} e^{-a_{ij}(t-s)}\left[\sum_{C_{kp}\in N_r(i,j)} C_{ij}^{kp} f(u_{kp}(s))u_{ij}(s) - v_{ij}(s)\right]ds. \tag{7.3}$$

Lemma 7.1.2. [13] *If* $u(t) \in \mathcal{B}$ *then* $\Pi u(t) \in \mathcal{B}$.

Proof. Fix a function $u(t) \in \mathcal{B}$. One can find that

$$
\begin{aligned}
\left|\Pi_{ij}u(t)\right| &\leqslant \int_{-\infty}^{t} e^{-a_{ij}(t-s)}\left(\sum_{C_{kp}\in N_r(i,j)} C_{ij}^{kp}|f(u_{kp})(s)u_{ij}(s)|+|v_{ij}(s)|\right)ds \\
&\leqslant \frac{1}{a_{ij}}\left(m_{ij}+Hm_f\sum_{C_{kp}\in N_r(i,j)} C_{ij}^{kp}\right),
\end{aligned} \tag{7.4}
$$

for all $i = 1, 2, \dots, m$, $j = 1, 2, \dots, n$. According to conditions (C6), (C7) we have that $\|\Pi u(t)\|_1 < H$.

Uniform continuity of function $\Pi u(t)$ follows estimates

$$
\begin{aligned}
\left|\frac{d\Pi_{ij}u(t)}{dt}\right| &= \mid \sum_{C_{kl}\in N_r(i,j)} C_{ij}^{kp}f(u_{kp}(t))u_{ij}(t) - v_{ij}(t) \\
&- \frac{1}{a_{ij}}\int_{-\infty}^{t} e^{-a_{ij}(t-s)}\left(\sum_{C_{kp}\in N_r(i,j)} C_{ij}^{kp}f(u_{kp})(s)u_{ij}(s)\right. \\
&\left. - v_{ij}(s)\right)ds\mid \leqslant \left(m_fH\sum_{C_{kp}\in N_r(i,j)} C_{ij}^{kp}+H\right) \\
&+\frac{1}{\gamma^2}\left(m_fH\sum_{C_{kp}\in N_r(i,j)} C_{ij}^{kp}+H\right) \\
&=\left(1+\frac{1}{\gamma^2}\right)H\left(m_f\sum_{C_{kp}\in N_r(i,j)} C_{ij}^{kp}+1\right),
\end{aligned}
$$

for all $i = 1, 2, \dots, m$, $j = 1, 2, \dots, n$. That is, the condition (K1) is valid since the derivatives are bounded.

Let us fix a positive number ϵ and a finite interval $[a, b] \subset \mathbb{R}$. Consider numbers $c < a$ and $\xi > 0$, which satisfy the following inequalities:

$$
\frac{2}{\gamma}\left(\max_{(i,j)}\sum_{C_{kp}\in N_r(i,j)} C_{ij}^{kp}(m_fH+LH^2)+m_{ij}\right)e^{-\gamma(a-c)} < \frac{\epsilon}{2}, \tag{7.5}
$$

$$
\frac{\xi}{\gamma}\left(\max_{(i,j)}\sum_{C_{kp}\in N_r(i,j)} C_{ij}^{kp}(m_f+LH)+1\right) < \frac{\epsilon}{2}. \tag{7.6}
$$

We will show that $\left\|\Pi u(t+t_p) - \Pi u(t)\right\| < \epsilon$ on $[a, b]$ for sufficiently large p. Let p be a large enough number such that $\left\|u(t+t_p) - u(t)\right\| < \xi$ and $\left\|v(t+t_p) - v(t)\right\| < \xi$, on $[c, b]$. Then for all $t \in [a, b]$ it is true that

$$
\begin{aligned}
&\left|\Pi_{ij}u(t+t_n)-\Pi_{ij}u(t)\right| \leqslant \\
&\int_{-\infty}^{c} e^{-\gamma(t-s)}\left(\sum_{C_{kp}\in N_r(i,j)} C_{ij}^{kp}\left|[f(u_{kp}(s))[u_{ij}(s)-u_{ij}(s+t_n)]\right.\right. \\
&\quad \left.\left.+[f(u_{kp}(s))-f(u_{kp}(s+t_n))]u_{ij}(s+t_n)]\right|+\left|(v_{ij}(s+t_n)-v_{ij}(s)\right|\right) ds \\
&+\int_{c}^{t} e^{-\gamma(t-s)}\left(\sum_{C_{kp}\in N_r(i,j)} C_{ij}^{kp}\left|f(u_{kp}(s))[u_{ij}(s)-u_{ij}(s+t_n)]\right.\right. \\
&\left.\left.+[f(u_{kp}(s))-f(u_{kp}(s+t_n))]u_{ij}(s+t_n)\right|+\left|(v_{ij}(s+t_n)-v_{ij}(s)\right|\right) ds \leqslant \\
&\frac{2}{\gamma}\left(\max_{(i,j)}\sum_{C_{kp}\in N_r(i,j)} C_{ij}^{kp}(m_f H+LH^2)+2m_{ij}\right)\times \\
&e^{-\gamma(a-c)}+\frac{\xi}{\gamma}\left(\max_{(i,j)}\sum_{C_{kp}\in N_r(i,j)} C_{ij}^{kp}(m_f+LH)+1\right),
\end{aligned}
$$

for all $i = 1, 2, \ldots, m$, $j = 1, 2, \ldots, n$. Now inequalities (2.4) and (2.5) imply that $\|\Pi u(t+t_n)-\Pi u(t)\| < \epsilon$ for $t \in [a, b]$. Since ϵ is an arbitrary small number the condition (K3) is valid. The lemma is proved. □

Lemma 7.1.3. [13] *The operator* Π *is contractive in* $\mathcal{B}$.

Proof. For two functions $\varphi, \psi \in \mathcal{B}$, we have that

$$
\begin{aligned}
&\left|\Pi_{ij}\varphi(t)-\Pi_{ij}\psi(t)\right| \\
&\leqslant \int_{-\infty}^{t} e^{-\gamma(t-s)} \sum_{C_{kp}\in N_r(i,j)} C_{ij}^{kp}\left|f(\varphi_{kp}(s))\varphi_{ij}(s)-f(\varphi_{kp}(s))\psi_{ij}(s)\right| ds \\
&+\int_{-\infty}^{t} e^{-\gamma(t-s)} \sum_{C_{kp}\in N_r(i,j)} C_{ij}^{kp}\left|f(\varphi_{kp}(s))\psi_{ij}(s)-f(\psi_{kp}(s))\psi_{ij}(s)\right| ds \\
&\leqslant \frac{(LH+m_f)}{\gamma}\max_{(i,j)}\sum_{C_{kp}\in N_r(i,j)} C_{ij}^{kp}\,\|\varphi-\psi\|_1 .
\end{aligned}
$$

This is why $\left\|\Pi\varphi-\Pi\psi\right\|_1 \leqslant \frac{(LH+m_f)}{\gamma}\max_{(i,j)}\sum_{C_{kp}\in N_r(i,j)} C_{ij}^{kp}\,\|\varphi-\psi\|_1$. Then condition (C8) implies that the operator Π is contractive in the set $\mathcal{B}$. The lemma is proved. □

Theorem 7.1.1. [13] *Suppose that conditions (C1)–(C8) are valid, then the system* (7.1) *possesses a unique asymptotically stable strongly unpredictable solution.*

Proof. Let us show that the space $\mathcal{B}$ is complete. Consider a Cauchy sequence $\phi_k(t)$ in $\mathcal{B}$, which converges to a limit function $\phi(t)$ on $\mathbb{R}$. It suffices to show that $\phi(t)$

satisfies condition (K3), since other two conditions can be easily checked. Fix a closed and bounded interval $I \subset \mathbb{R}$. We have that

$$\begin{aligned}&\|\phi(t + t_p) - \phi(t)\| \leqslant \|\phi(t + t_p) - \phi_k(t + t_p)\|\\&+\|\phi_k(t + t_p) - \phi_k(t)\| + \|\phi_k(t) - \phi(t)\|.\end{aligned} \tag{7.7}$$

Now, one can take sufficiently large p and k such that each term on the right-hand side of equation (7.31) is smaller than $\frac{\epsilon}{3}$ for an arbitrary positive ϵ and $t \in I$. The inequality implies that $\|\phi(t + t_p) - \phi(t)\| \leqslant \epsilon$ on I. That is the sequence $\phi(t + t_p)$ uniformly converges to $\phi(t)$ on I. The completeness of $\mathcal{B}$ is proved. Now, by the contractive mapping theorem, due to lemmas 7.1.2 and 7.1.3, there exists a unique solution $\omega(t) \in \mathcal{B}$ of the equation (7.1).

Next, using the relations

$$\begin{aligned}&\omega_{ij}(t) = \omega_{ij}(s_p) - \int_{s_p}^{t} a_{ij}\omega_{ij}(s)ds\\&-\int_{s_p}^{t} \sum_{C_{kl}\in N_r(i,j)} C_{ij}^{kp} f(\omega_{kp}(s))\omega_{ij}(s)ds + \int_{s_p}^{t} v_{ij}(s)ds,\end{aligned}$$

and

$$\begin{aligned}&\omega_{ij}(t + t_p) = \omega_{ij}(t_p + s_p) - \int_{s_p}^{t} a_{ij}\omega_{ij}(s + t_p)ds\\&-\int_{s_p}^{t} \sum_{C_{kp}\in N_r(i,j)} C_{ij}^{kp} f(\omega_{kp}(s + t_p))\omega_{ij}(s + t_p)ds\\&+\int_{s_s}^{t} v_{ij}(s + t_p)ds,\end{aligned}$$

we obtain that

$$\begin{aligned}&\omega_{ij}(t + t_p) - \omega_{ij}(t) = \omega(t_p + s_p) - \omega(s_p)\\&-\int_{s_p}^{t} a_{ij}(\omega(s + t_p) - \omega(s))ds\\&-\int_{s_p}^{t} (v_{ij}(s + t_p) - v_{ij}(s))ds\\&-\int_{s_p}^{t} \sum_{C_{kp}\in N_r(i,j)} C_{ij}^{kp}(f(\omega_{kp}(s + t_p))\omega_{ij}(s + t_p)\\&-f(\omega_{kp}(s))\omega_{ij}(s))ds.\end{aligned} \tag{7.8}$$

There exist a positive number κ and an integers l and k such that the following inequalities are valid:

$$\kappa < \delta; \tag{7.9}$$

$$|\omega_{ij}(t+s)-\omega_{ij}(t)|<\frac{\epsilon_0}{k}, \quad t\in\mathbb{R}, |s|<\kappa; \tag{7.10}$$

$$\frac{1}{k}(\bar{\gamma}+\sum_{C_{kp}\in N_r(i,j)} C_{ij}^{kp}(m_f+LH))<1; \tag{7.11}$$

$$\kappa\left(1-\frac{1}{k}(\bar{\gamma}+\sum_{C_{kp}\in N_r(i,j)} C_{ij}^{kp}(m_f+LH))\right)\geqslant\frac{3}{2l}, \tag{7.12}$$

where $\bar{\gamma}=\max_{(i,j)}a_{ij}$, for all $i=1, 2, \ldots, m, j=1, 2, \ldots, n$.

Let the numbers κ, l and k as well as numbers $p\in\mathbb{N}$, and $i=1, 2, \ldots, m, j=1, 2, \ldots, n$, be fixed. Denote $\Delta=|\omega_{ij}(t_p+s_p)-\omega_{ij}(s_p)|$. Consider the following two alternatives, (i) $\Delta<\epsilon_0/l$ and (ii) $\Delta\geqslant\epsilon_0/l$, such that the remaining proof falls naturally into two parts.

(i) It is true that

$$|v_{ij}(t+s_p)-v_{ij}(s_p)|>\epsilon_0,$$

if $t\in[s_p-\delta, s_p+\delta]$.

Therefore, by using relations (7.8)–(7.4) we have that

$$\begin{aligned}
|\omega_{ij}(t+t_p)-\omega_{ij}(t)| &\geqslant \int_{s_p}^{t}|v_{ij}(s+t_p)-v_{ij}(s)|ds\\
&-|\omega_{ij}(t_p+s_p)-\omega_{ij}(s_p)|-\int_{s_p}^{t}a_{ij}|\omega_{ij}(s+t_p)-\omega_{ij}(s)|ds\\
&-\int_{s_p}^{t}\sum_{C_{kp}\in N_r(i,j)} C_{ij}^{kp}|f(\omega_{kp}(s+t_p))\omega_{ij}(s+t_p)-f(\omega_{kp}(s))\omega_{ij}(s)|ds\\
&\geqslant\epsilon_0\kappa-\frac{\epsilon_0}{l}-\frac{\epsilon_0}{k}\kappa(\bar{\gamma}+\sum_{C_{kp}\in N_r(i,j)} C_{ij}^{kp}(m_f+LH))\\
&=\epsilon_0\kappa\left(1-\frac{1}{k}(\bar{\gamma}+\sum_{C_{kp}\in N_r(i,j)} C_{ij}^{kp}(m_f+LH))\right)-\frac{\epsilon_0}{l}\geqslant\frac{\epsilon_0}{2l},
\end{aligned}$$

for $t\in[s_p-\kappa, s_p+\kappa]$.

(ii) Consider the case $\Delta\geqslant\epsilon_0/l$. There exists a positive number $\kappa_1\leqslant\kappa$ such that

$$|\omega_{ij}(t+s)-\omega_{ij}(t)|<\frac{\epsilon_0}{4l}, \quad t\in\mathbb{R}, |s|<\kappa_1.$$

Therefore, we obtain that

$$\begin{aligned}
|\omega_{ij}(t_p+t)-\omega_{ij}(t)| &\geqslant |\omega_{ij}(t_p+s_p)-\omega_{ij}(s_p)|\\
&-|\omega_{ij}(s_p)-\omega_{ij}(t)|-|\omega_{ij}(t_p+t)-\omega_{ij}(t_p+s_p)|\\
&\geqslant\frac{\epsilon_0}{l}-\frac{\epsilon_0}{4l}-\frac{\epsilon_0}{4l}=\frac{\epsilon_0}{2l},
\end{aligned}$$

for $t \in [s_p - \kappa_1, s_p + \kappa_1]$ and $i = 1, 2, \ldots, m, j = 1, 2, \ldots, n$.

Thus one can conclude that $\omega(t)$ is a strongly unpredictable solution with positive numbers $\frac{\epsilon_0}{2l}$, κ_1, and the sequences t_p and u_p.

Finally, we will discuss the stability of the solution $\omega(t)$. It is true that

$$\omega_{ij}(t) = e^{-a_{ij}(t-t_0)}\omega_{ij}(t_0) - \int_{t_0}^{t} e^{-a_{ij}(t-s)}\left[\sum_{C_{kp}\in N_r(i,j)} C_{ij}^{kp} f(\omega_{kp}(s))\omega_{ij}(s) - v_{ij(s)}\right]ds,$$

$i = 1, \ldots, m, j = 1, \ldots, n$.

Let $z(t) = (z_{11}, z_{12}, \ldots, z_{mn})$ be another solution of system (7.1). One can write

$$z_{ij}(t) = e^{-a_{ij}(t-t_0)}z_{ij}(t_0) - \int_{t_0}^{t} e^{-a_{ij}(t-s)}\left[\sum_{C_{kp}\in N_r(i,j)} C_{ij}^{kp} f(z_{kp}(s))z_{ij}(s) - v_{ij(s)}\right]ds$$

for all $i = 1, \ldots, m, j = 1, \ldots, n$. Making use of the relation

$$z_{ij}(t) - \omega_{ij}(t) = e^{-a_{ij}(t-t_0)}\big(z_{ij}(t_0) - \omega_{ij}(t_0)\big) - \int_{t_0}^{t} e^{-a_{ij}(t-s)}\left[\sum_{C_{kp}\in N_r(i,j)} C_{ij}^{kp} f(z_{kp}(s))z_{ij}(s) - \sum_{C_{kp}\in N_r(i,j)} C_{ij}^{kp} f(\omega_{kp}(s))\omega_{ij}(s)\right]ds,$$

we obtain that

$$\begin{aligned}
\left|z_{ij}(t) - \omega_{ij}(t)\right| &\leqslant e^{-\gamma(t-t_0)}\left|z_{ij}(t_0) - \omega_{ij}(t_0)\right| \\
&+ m_f \sum_{C_{kp}\in N_r(i,j)} C_{ij}^{kp} \int_{t_0}^{t} e^{-\gamma(t-s)}\left|z_{ij}(s) - \omega_{ij}(s)\right| ds \\
&+ LH \int_{t_0}^{t} e^{-\gamma(t-s)} \sum_{C_{kp}\in N_r(i,j)} C_{ij}^{kp}\left|z_{kp}(s) - \omega_{kp}(s)\right| ds \\
&\leqslant e^{-\gamma(t-t_0)}\|z(t_0) - \omega(t_0)\| \\
&+ m_f \sum_{C_{kp}\in N_r(i,j)} C_{ij}^{kp} \int_{t_0}^{t} e^{-\gamma(t-s)}\|z(s) - \omega(s)\|ds \\
&+ LH \sum_{C_{kp}\in N_r(i,j)} C_{ij}^{kp} \int_{t_0}^{t} e^{-\gamma(t-s)}\|z(s) - \omega(s)\|ds,
\end{aligned}$$

for all $i = 1, 2, \ldots, m, j = 1, 2, \ldots, n$.

Thus it can be confirmed that

$$\|z(t) - \omega(t)\| \leqslant e^{-\gamma(t-t_0)}\|z(t_0) - \omega(t_0)\| + D\int_{t_0}^{t} e^{-\gamma(t-s)}\|z(s) - \omega(s)\|ds,$$

where $D = (LH + m_f)\sum_{C_{kp}\in N_r(i,j)} C_{ij}^{kp}$. Multiplying both sides of the last inequality by $e^{\gamma t}$ we obtain that

$$e^{\gamma t} \parallel z(t) - \omega(t)\parallel \leqslant e^{\gamma t_0} \parallel z(t_0) - \omega(t_0)\parallel$$
$$+ D\int_{t_0}^{t} e^{\gamma s} \parallel z(s) - \omega(s)\parallel ds.$$

Now, applying the Gronwall–Bellman lemma, one can obtain that

$$\parallel z(t) - \omega(t)\parallel \leqslant \parallel z(t_0) - \omega(t_0)\parallel e^{[D-\gamma](t-t_0)}.$$

The last inequality and condition (C8) confirm that the strongly unpredictable solution $\omega(t)$ is uniformly asymptotically stable. The theorem is proved. □

7.1.3 Examples

Example 7.1.1. [13] *It was shown in example* 2.2.1 *that the logistic map*

$$\lambda_{i+1} = \mu\lambda_i(1 - \lambda_i), \tag{7.13}$$

possesses an unpredictable solution for each $\mu \in [3 + (2/3)^{1/2}, 4]$. *Let* ψ_i, $i \in \mathbb{Z}$, *be the unpredictable solution of the equation with* $\mu = 3.92$. *The sequence belongs to the unit interval* [0, 1]. *There exist a positive number* ϵ_0 *and sequences* ζ_p, η_p, *both of which diverge to infinity such that* $|\psi_{i+\zeta_p} - \psi_i| \to 0$ *as* $p \to \infty$ *for each i in bounded intervals of integers and* $|\psi_{\zeta_p+\eta_p} - \psi_{\eta_p}| \geqslant \epsilon_0$ *for each* $p \in \mathbb{N}$.

Next, we will consider the unique unpredictable solution, $\Theta(t)$, *of the equation*

$$\frac{dv}{dt} = -3v(t) + \Omega(t), \tag{7.14}$$

where $\Omega(t)$ *is a piecewise constant function defined on the real axis through the equation* $\Omega(t) = \psi_i$ *for* $t \in [i, i + 1)$, $i \in \mathbb{Z}$. *One can check that*

$$\Theta(t) = \int_{-\infty}^{t} e^{-3(t-s)}\Omega(s)ds. \tag{7.15}$$

It is worth noting that $\Theta(t)$ *is bounded on the whole real axis such that* $\sup_{t\in\mathbb{R}} |\Theta(t)| \leqslant 1/3$, *and is globally exponentially stable. Identically with that considered in example* 6.4.1, *we can show that the function* $\Theta(t)$ *is unpredictable. Figure* 7.1 *presents the irregular behavior of the function by simulation of a solution of equation* (7.14), *which is attracted by the unpredictable solution.*

Example 7.1.2. [13] *Let us introduce the following SICNNs*:

$$\frac{dx_{ij}}{dt} = -a_{ij}x_{ij} - \sum_{C_{kp}\in N_1(i,j)} C_{ij}^{kp} f(x_{kp}(t))x_{ij} + v_{ij}(t), \tag{7.16}$$

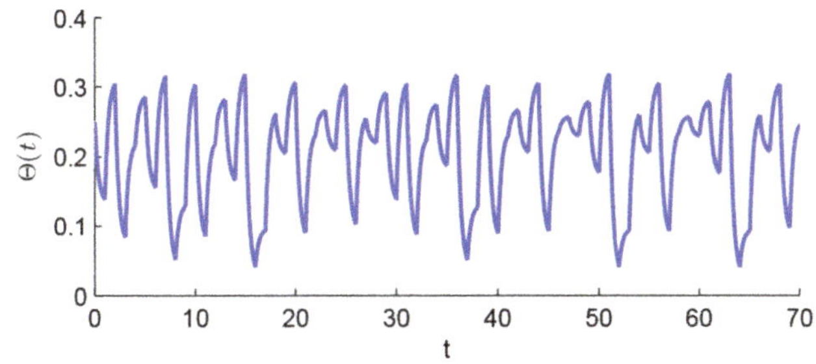

Figure 7.1. The simulation of a solution of equation (7.14), which is attracted by the unpredictable solution $\Theta(t)$. Reprinted from [13], Copyright (2020), with permission from Elsevier.

where $i, j = 1, 2, 3$,

$$\begin{pmatrix} a_{11} & a_{12} & a_{13} \\ a_{21} & a_{22} & a_{23} \\ a_{31} & a_{32} & a_{33} \end{pmatrix} = \begin{pmatrix} 3 & 8 & 5 \\ 4 & 9 & 6 \\ 2 & 7 & 3 \end{pmatrix},$$

$$\begin{pmatrix} C_{11} & C_{12} & C_{13} \\ C_{21} & C_{22} & C_{23} \\ C_{31} & C_{32} & C_{33} \end{pmatrix} = \begin{pmatrix} 0.004 & 0.003 & 0 \\ 0.007 & 0.006 & 0.002 \\ 0.005 & 0 & 0.008 \end{pmatrix},$$

and $f(s) = \arctan(s)$, $v_{11}(t) = 25\Theta^3(t) + 1$, $v_{12}(t) = 4\Theta(t)$, $v_{13}(t) = -3\Theta(t) + 2$, $v_{21}(t) = 15\Theta(t) + 3$, $v_{22}(t) = 17\Theta^3(t)$, $v_{23}(t) = 18\Theta(t) - 1$, $v_{31}(t) = -7\Theta(t) + 2$, $v_{32}(t) = 3\Theta(t)$ and $v_{33}(t) = -13\Theta^3(t)$, *where* $\Theta(t)$ *is the unpredictable solution of system* (7.14). *One can confirm that* $|v_{ij}(t)| \leqslant m_{ij}$, *where* $m_{11} = 1.93$; $m_{12} = 1.34$, $m_{13} = 3$, $m_{21} = 8$, $m_{22} = 0.63$, $m_{23} = 7$, $m_{31} = 4.34$, $m_{32} = 1$ and $m_{33} = 0.49$. *Moreover, by means of lemmas* 1.4 *and* 1.5 *in* [8], *functions* $v_{ij}(t)$, $i = 1, 2, 3$, $j = 1, 2, 3$, *are unpredictable.*

Figure 7.2 *represents the solution,* $\phi(t)$, *of* (7.16), *which is attracted by the unpredictable solution, and consequently illustrates it. The initial values of* $\phi(t)$ *are* $\phi_{11}(0) = 0.524\,5$, $\phi_{12}(0) = 0.149\,6$, $\phi_{13}(0) = 0.223\,7$, $\phi_{21}(0) = 1.828\,3$, $\phi_{22}(0) = 0.051\,3$, $\phi_{23}(0) = 0.772\,5$, $\phi_{31}(0) = 0.067\,8$, $\phi_{32}(0) = 0.127\,8$ and $\phi_{33}(0) = -0.099\,9$.

In the following example we take the coordinates of the solution $\phi(t)$ of system (7.16) as perturbation functions.

Example 7.1.3. [13] *Consider the following SICNNs:*

$$\frac{dy_{ij}}{dt} = -a_{ij}y_{ij} - \sum_{C_{kp}\in N_1(i,j)} C_{ij}^{kp} f(y_{kp}(t))y_{ij} + v_{ij}(t), \tag{7.17}$$

$i, j = 1, 2, 3$, *where* $f(s) = 0.5s^3$, $v_{11}(t) = 3\phi_{12}(t)$, $v_{12}(t) = -\Theta^3(t)$, $v_{13}(t) = -4\phi_{13}(t)$, $v_{21}(t) = 8\phi_{32}(t)$, $v_{22}(t) = -6\phi_{33}(t)$, $v_{23}(t) = 5\phi_{21}(t)$, $v_{31}(t) = 4\phi_{13}^3(t)$, $v_{32}(t) = 2\Theta^3(t)$ and $v_{33}(t) = 9\phi_{31}(t)$,

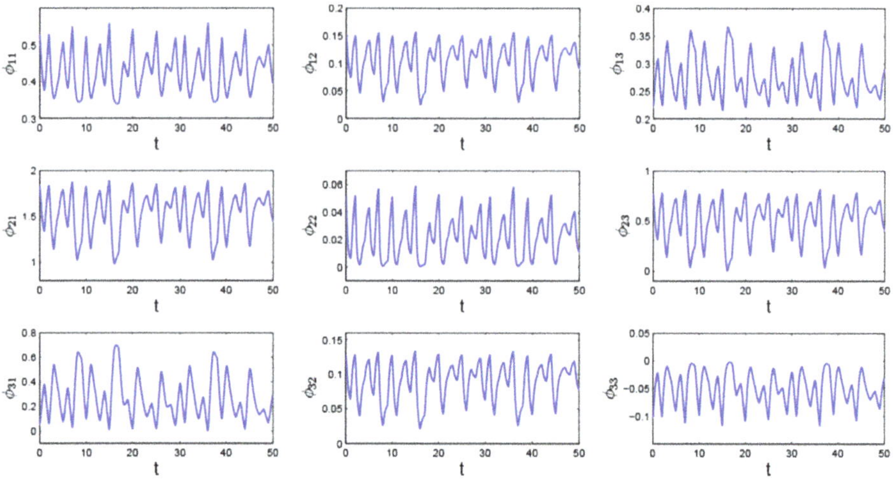

Figure 7.2. The coordinates of the solution $\phi(t)$ of the SICNNs (7.16). Reprinted from [13], Copyright (2020), with permission from Elsevier.

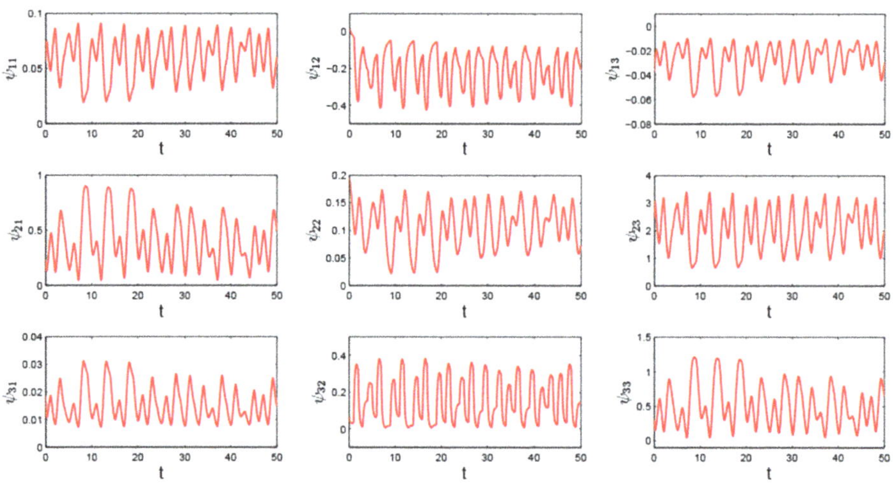

Figure 7.3. The coordinates of the solution $\psi(t)$ of the SICNNs (7.17).

$$\begin{pmatrix} a_{11} & a_{12} & a_{13} \\ a_{21} & a_{22} & a_{23} \\ a_{31} & a_{32} & a_{33} \end{pmatrix} = \begin{pmatrix} 5 & 1 & 3 \\ 6 & 2 & 8 \\ 6 & 4 & 5 \end{pmatrix},$$

$$\begin{pmatrix} C_{11} & C_{12} & C_{13} \\ C_{21} & C_{22} & C_{23} \\ C_{31} & C_{32} & C_{33} \end{pmatrix} = \begin{pmatrix} 0.02 & 0 & 0.06 \\ 0.04 & 0.07 & 0.09 \\ 0.03 & 0.04 & 0 \end{pmatrix}.$$

Figure 7.3 *reveals the result of theorem* 7.1.8 *and depicts the solution* $\psi(t)$ *of system* (7.17)*, which is attracted by the unpredictable solution, and we are confident that the simulation illustrates the result of the section.*

The solution $\psi(t)$ *has initial values* $\psi_{11}(0) = 0.124\,7$, $\psi_{12}(0) = 0.019\,3$, $\psi_{13}(0) = 0.004\,3$, $\psi_{21}(0) = 0.560\,3$, $\psi_{22}(0) = 0.189\,4$, $\psi_{23}(0) = 2.457\,4$, $\psi_{31}(0) = 0.006\,7$, $\psi_{32}(0) = 0.064\,5$ and $\psi_{33}(0) = 0.011\,3$.

7.2 Inertial neural networks with unpredictable oscillations

In this section a new type of oscillation for inertial neural networks is investigated. Unpredictable solutions that have chaotic behavior are discussed. The existence and uniqueness theorem are proved and asymptotic stability is investigated. Numerical simulations are presented to demonstrate the validity of the theoretical results.

In recent decades dynamicists have focused on the study neural networks. Many results on stability, periodicity, synchronization analysis and chaos in neural networks have been published. Some authors have investigated neural networks by adding inertia. For example, inertial bidirectional associative memory (BAM) neural networks [38, 39, 41, 54, 72, 73], inertial Cohen–Grossberg-type neural networks [40, 71], electronic neural networks with inertia [19] and inertial memristive neural networks [55, 58, 59] have been studied.

7.2.1 The model

In this section we consider unpredictable oscillations which are described by unpredictable functions, which were introduced in [4]. The unpredictable oscillations continue a line of periodic and almost periodic motions of neural networks. The existence of unpredictable solutions proves Poincaré chaos for a Hopfield type neural network [4] and shunting inhibitory cellular neural networks have been proved [13]. The motions admit numeric characteristics, which can be useful for the analysis of neural processes similarly to the way that periodic and almost periodic oscillations are. The description of such functions relies on the dynamics of unpredictable points [4–6]. The role of unpredictable points is very significant in the study of chaos and this is confirmed by the works [51] and [63], where Poincaré, strongly Ruelle–Takens and strongly Auslander–Yorke chaos are considered by using the unpredictable points. The introduced unpredictable functions lead to the investigation of chaos to differential and discrete equations [11, 12]. The research of these functions combines theoretical advantages and problems, which are suitable for both oscillations and chaos, and will open up many interesting prospects in neuroscience.

Our main purpose is to give the conditions ensuring the existence, uniqueness and asymptotic stability of the unpredictable oscillations in inertial neural networks.

Throughout this chapter, $\mathbb{R}$ and $\mathbb{N}$ will represent the sets of real and natural numbers, respectively. Additionally, the norm $\|v\|_1 = \sup_{t\in\mathbb{R}}\|v(t)\|$, where $\|v\| = \max_{1\leqslant i\leqslant p}|v_i|$, $v = (v_1, \dots, v_p)$, $t, v_i \in \mathbb{R}$, $i = 1, \dots, p$, will be used. The following definition is one of the main definitions in our study.

Definition 7.2.1. [4] *A uniformly continuous and bounded function* $v: \mathbb{R} \to \mathbb{R}^p$ *is unpredictable if there exist positive numbers* ϵ_0, δ *and sequences* t_n, u_n *both of which* *diverge to infinity such that $v(t + t_n) \to v(t)$ as $n \to \infty$ uniformly on compact subsets of $\mathbb{R}$ and $\|v(t + t_n) - v(t)\| \geqslant \epsilon_0$ for each $t \in [u_n - \delta, u_n + \delta]$ and $n \in \mathbb{N}$.*

In this chapter we consider the following inertial neural networks:

$$\frac{d^2x_i(t)}{dt^2} = -a_i\frac{x_i(t)}{dt} - b_ix_i(t) + \sum_{j=1}^{p} c_{ij}f_j(x_j(t)) + v_i(t), \tag{7.18}$$

where $t, x_i \in \mathbb{R}$, $i = 1, \ldots p$, p denotes the number of neurons in the network, $x_i(t)$ with $i = 1, \ldots p$ corresponds to the state of the unit i at time t, the second derivative is called an inertial term, $b_i > 0$, $a_i > 0$ are constants, f_i with $i = 1, \ldots p$ denote the measures of activation to its incoming potentials of the unit i at time t, c_{ij} for all with $i, j = 1, \ldots p$ are constants, which denote the synaptic connection weight of the unit j on the unit i, and $v_i(t)$ are external inputs on the ith neuron at the time t. We assume that the coefficients c_{ij} are real and the activation functions $f_i: \mathbb{R} \to \mathbb{R}$ are continuous functions satisfying the following condition:

1. $|f_i(x_1) - f_i(x_2)| \leqslant L_i|x_1 - x_2|$ for all $x_1, x_2 \in \mathbb{R}$, where $L_i > 0$ are Lipschitz constants, for $i = 1, \ldots, p$ and $\max_{1\leqslant i\leqslant p}L_i = L$.

Under variable transformation

$$y_i(t) = \xi_i\frac{dx_i(t)}{dt} + \zeta_i x_i(t),\ i = 1, \ldots, p, \tag{7.19}$$

the system (7.18) can be rewritten as

$$\frac{dx_i(t)}{dt} = -\frac{\zeta_i}{\xi_i}x_i(t) + \frac{1}{\xi_i}y_i(t), \tag{7.20}$$

$$\begin{aligned}\frac{dy_i(t)}{dt} = &-\left(\xi_i b_i - \zeta_i\left(a_i - \frac{\zeta_i}{\xi_i}\right)\right)x_i(t) - \left(a_i - \frac{\zeta_i}{\xi_i}\right)y_i(t)\\ &+ \xi_i\sum_{j=1}^{p} c_{ij}f_j(x_j(t)) + \xi_i v_i(t).\end{aligned} \tag{7.21}$$

According to the results in [26] $x(t) = (x_1(t), x_2(t), \ldots, x_p(t))$ and $y(t) = (y_1(t), y_2(t), \ldots, y_p(t))$ is a bounded solution of equations (7.20), (7.21) if and only if the following integral equations are satisfied:

$$x_i(t) = \frac{1}{\xi_i}\int_{-\infty}^{t} e^{-\frac{\zeta_i}{\xi_i}(t-s)}y_i(s)ds, \tag{7.22}$$

$$y_i(t) = \int_{-\infty}^{t} e^{-(a_i-\frac{\zeta_i}{\xi_i})(t-s)}\left[(\zeta_i(a_i - \frac{\zeta_i}{\xi_i}) - \xi_i b_i)x_i(s) + \xi_i\sum_{j=1}^{p} c_{ij}f_j(x_j(s)) + \xi_i v_i(s)\right]ds. \tag{7.23}$$

Let us define a $2p$-dimensional function $z(x, y)$ and rewrite systems (7.20) and (7.21):

$$\frac{dz(t)}{dt} = Az(t) + F(z(t)), \tag{7.24}$$

where the matrix

$$A = \text{diag}\left\{-\frac{\zeta_1}{\xi_1}, \ldots, -\frac{\zeta_p}{\xi_p}, -\left(a_1 - \frac{\zeta_1}{\xi_1}\right), \ldots, -\left(a_p - \frac{\zeta_p}{\xi_p}\right)\right\},$$

and vector function $F(z(t)) = (F_1(z(t)), F_2(z(t)), \ldots, F_{2p}(z(t)))$, such that

$$F_i(z(t)) = \begin{cases} \frac{1}{\xi_i} z_{i+p}(t) & i = 1, \ldots, p, \\ -\left(\xi_{i-p}b_{i-p} - \zeta_{i-p}\left(a_{i-p} - \frac{\zeta_{i-p}}{\xi_{i-p}}\right)\right)z_{i-p}(t) + \xi_{i-p}\sum_{j=1}^{p} c_{(i-p)j} & i = p+1, \ldots, 2p. \\ f_j(z_j(t)) + \xi_{i-p}v_{i-p}(t) & \end{cases}$$

7.2.2 The main result

Denote by Σ the space of $2p$-dimensional vector functions, $\varphi\colon \mathbb{R} \to \mathbb{R}^{2p}$, $\varphi = (\varphi_1, \ldots, \varphi_{2p})$, $p-$ is a natural number, such that:

1. Function $\phi(t)$ is uniformly continuous.
2. There exists a positive number H such that $\|\phi\|_1 < H$ for all $\phi(t) \in \Sigma$.
3. There exists a sequence t_n, $t_n \to \infty$ as $n \to \infty$ such that for each $\phi(t) \in \Sigma$ the sequence $\phi(t + t_n)$ uniformly converges to $\phi(t)$ on each closed and bounded interval of the real axis.

Let us introduce the operator $\Pi\colon \Sigma \to \Sigma$, such that

$$\Pi_i\phi_i(t) = \begin{cases} \frac{1}{\xi_i}\int_{-\infty}^{t} e^{-\frac{\zeta_i}{\xi_i}(t-s)}\phi_{i+p}(s)ds, & i = 1, \ldots, p, \\ \int_{-\infty}^{t} e^{-\left(a_{i-p}-\frac{\zeta_{i-p}}{\xi_{i-p}}\right)(t-s)}\left[\left(\zeta_{i-p}\left(a_{i-p} - \frac{\zeta_{i-p}}{\xi_{i-p}}\right) - \xi_{i-p}b_{i-p}\right)\right. & \\ \phi_{i-p}(s) & \\ \left. + \xi_{i-p}\sum_{j=1}^{p} c_{(i-p)j}f_j(\phi_j(s)) + \xi_{i-p}v_{i-p}(s)\right]ds, & i = p+1, \ldots, 2p. \end{cases} \tag{7.25}$$

The following conditions will be needed throughout the chapter:

1. The functions $v_i(t)$, $i = 1, \ldots, p$, in system (7.18) are unpredictable, they belong to Σ and there exist positive numbers δ, $\epsilon_0 > 0$ and a sequence $s_n \to \infty$ as $p \to \infty$, which satisfy $|v_i(t + t_n) - v_i(t)| \geqslant \epsilon_0$ for all $t \in [s_n - \delta, s_n + \delta]$, $i = 1, \ldots, 2p$, and $n \in \mathbb{N}$.
2. There exists a positive number M_f such that $|f_i(s)| \leqslant M_f$, for $|s| < H$ and $i = 1, 2, \ldots, p$.

There exist positive real numbers ξ_i, ζ_i, such that the following assumptions are satisfied:

1. $a_i > \frac{\zeta_i}{\xi_i} > 0$, $\zeta_i > \xi_i > 1$, $i = 1, \dots, p$.
2. $\left(a_i - \frac{\zeta_i}{\xi_i}\right) - \left(\left|\zeta_i\left(a_i - \frac{\zeta_i}{\xi_i}\right) - \xi_i b_i\right| + \xi_i\right) > 0$, for each $i = 1, \dots, p$.
3. $\dfrac{\xi_i M_f \sum_{j=1}^p c_{ij}}{\left(a_i - \frac{\zeta_i}{\xi_i}\right) - \left(\left|\zeta_i\left(a_i - \frac{\zeta_i}{\xi_i}\right) - \xi_i b_i\right| + \xi_i\right)} < H$, $i = 1, \dots, p$.
4. $\dfrac{1}{\left(a_i - \frac{\zeta_i}{\xi_i}\right)}\left[\left|\zeta_i\left(a_i - \frac{\zeta_i}{\xi_i}\right) - \xi_i b_i\right| + L\xi_i \sum_{j=1}^p c_{ij}\right] < 1$, $i = 1, \dots, p$.
5. $\max\left(\frac{1}{\xi_i}, \left|-\xi_i b_i + \zeta_i\left(a_i - \frac{\zeta_i}{\xi_i}\right)\right| + L\xi_i \sum_{j=1}^p c_{ij}\right) < \min\left(\frac{\zeta_i}{\xi_i}, a_i - \frac{\zeta_i}{\xi_i}\right)$, $i = 1, \dots, p$.

Lemma 7.2.1. [16] *If the function* $\phi(t) \in \Sigma$ *then* $\Pi\phi(t) \in \Sigma$.

Proof. Uniform continuity of function $\Pi\varphi$ follows from the boundedness of its derivative.

For function $\phi(t) \in \Sigma$ we have

$$
|\Pi_i\phi_i(t)| = \begin{cases}
\left|\dfrac{1}{\xi_i}\displaystyle\int_{-\infty}^{t} e^{-\frac{\zeta_i}{\xi_i}(t-s)}\phi_{i+p}(s)ds\right| \leqslant \dfrac{1}{\zeta_i}|\phi_{i+p}(t)| \leqslant \dfrac{H}{\zeta_i} < H, \\
i = 1, \dots, p, \\
\left|\displaystyle\int_{-\infty}^{t} e^{-(a_{i-p}-\frac{\zeta_{i-p}}{\xi_{i-p}})(t-s)}\left[\left(\zeta_{i-p}\left(a_{i-p} - \dfrac{\zeta_{i-p}}{\xi_{i-p}}\right) - \xi_{i-p}b_{i-p}\right)\phi_{i-p}(s)\right.\right. \\
\left.\left. + \xi_{i-p}\displaystyle\sum_{j=1}^{p} c_{(i-p)j}f_j(\phi_j(s)) + \xi_{i-p}v_{i-p}(s)\right]ds\right| \\
\leqslant \displaystyle\int_{-\infty}^{t} e^{-\left(a_{i-p}-\frac{\zeta_{i-p}}{\xi_{i-p}}\right)(t-s)} \\
\left[\left|\zeta_{i-p}\left(a_{i-p} - \dfrac{\zeta_{i-p}}{\xi_{i-p}}\right) - \xi_{i-p}b_{i-p}\right|H + \xi_{i-p}\displaystyle\sum_{j=1}^{p} c_{(i-p)j}M_f + \xi_{i-p}H\right]ds \\
\leqslant \dfrac{1}{a_{i-p} - \dfrac{\zeta_{i-p}}{\xi_{i-p}}} \\
\left[H\left(\left|\zeta_{i-p}\left(a_{i-p} - \dfrac{\zeta_{i-p}}{\xi_{i-p}}\right) - \xi_{i-p}b_{i-p}\right| + \xi_{i-p}\right) + \xi_{i-p}M_f\displaystyle\sum_{j=1}^{p} c_{(i-p)j}\right], \\
i = p + 1, \dots, 2p.
\end{cases}
$$

Conditions (C5) and (C6) imply that $|\Pi_i\phi_i(t)| < H$, for each $i = 1, \dots, 2p$. So that $\|\Pi\phi\|_1 = \max_i|\Pi_i\phi_i| < H$. Thus condition (K2) is valid.

For the function $\phi(t)$ belonging to Σ and fixed $i = 1, \dots, 2p$, we have that

$$|\Pi_i\phi_i(t + t_n) - \Pi_i\phi_i(t) = \begin{cases} \left|\dfrac{1}{\xi_i}\displaystyle\int_{-\infty}^{t} e^{-\frac{\zeta_i}{\xi_i}(t-s)}(\phi_{i+p}(s + t_n) - \phi_{i+p}(s))ds\right|, \\ i = 1, \dots, p, \\ \Bigg|\displaystyle\int_{-\infty}^{t} e^{-\left(a_{i-p}-\frac{\xi_{i-p}}{\zeta_{i-p}}\right)(t-s)} \\ \left[\left(\zeta_{i-p}\left(a_{i-p} - \dfrac{\zeta_{i-p}}{\xi_{i-p}}\right) - \xi_{i-p}b_{i-p}\right)(\phi_{i-p}(s + t_n) - \phi_{i-p}(s))\right. \\ + \xi_{i-p}\displaystyle\sum_{j=1}^{p} c_{(i-p)j} \\ \left.(f_j(\phi_j(s + t_n) - f_j(\phi_j(s))) + \xi_{i-p}(v_{i-p}(s + t_n) - v_{i-p}(s))\right]ds\Bigg|, \\ i = p + 1, \dots, 2p. \end{cases}$$

Let us fix a positive number ϵ and a section $[a, b]$, $-\infty < a < b < \infty$. We will show that for sufficiently large n it is true that $\|\Pi\phi(t + t_n) - \Pi\phi(t)\| < \epsilon$ on $[a, b]$. Choose numbers $c < a$ and $0 < \xi$, satisfying the following inequalities:

$$\frac{2H}{\zeta_i}e^{-\frac{\zeta_i}{\xi_i}(a-c)} < \frac{\epsilon}{2}, \tag{7.26}$$

$$\frac{2H}{a_i - \frac{\zeta_i}{\xi_i}}\left(|\zeta_i\left(a_i - \frac{\zeta_i}{\xi_i}\right) - \xi_i b_i| + L\xi_i\sum_{j=1}^{p} c_{ij} + \xi_i\right)e^{-\left(a_i-\frac{\zeta_i}{\xi_i}\right)(a-c)} < \frac{\epsilon}{2}, \tag{7.27}$$

$$\frac{\xi}{\zeta_i} < \frac{\epsilon}{2}, \tag{7.28}$$

and

$$\frac{\xi}{a_i - \frac{\zeta_i}{\xi_i}}\left(|\zeta_i\left(a_i - \frac{\zeta_i}{\xi_i}\right) - \xi_i b_i| + L\xi_i\sum_{j=1}^{p} c_{ij} + \xi_i\right) < \frac{\epsilon}{2}. \tag{7.29}$$

Consider the number n sufficiently large such that $|\phi_i(t + t_n) - \phi_i(t)| < \xi$, $i = 1, \dots, 2p$, and $|v_i(t + t_n) - v_i(t)| < \xi$, $i = 1, \dots, p$, on $[c, b]$. Then, for all $t \in [a, b]$, it is true that

$$
|\Pi_i\phi_i(t+t_n)-\Pi_i\phi_i(t)| \leqslant
$$

$$
\leqslant \begin{cases}
\left|\dfrac{1}{\xi_i}\displaystyle\int_{-\infty}^{c} e^{-\frac{\zeta_i}{\xi_i}(t-s)}(\phi_{i+p}(s+t_n)-\phi_{i+p}(s))ds\right| \\
+\left|\dfrac{1}{\xi_i}\displaystyle\int_{c}^{t} e^{-\frac{\zeta_i}{\xi_i}(t-s)}(\phi_{i+p}(s+t_n)-\phi_{i+p}(s))ds\right| \leqslant \dfrac{2H}{\zeta_i}e^{-\frac{\zeta_i}{\xi_i}(a-c)}+\dfrac{1}{\zeta_i}\xi<\epsilon, \\
i=1,\ldots,p, \\
\left|\displaystyle\int_{-\infty}^{c} e^{-\left(a_{i-p}-\frac{\zeta_{i-p}}{\xi_{i-p}}\right)(t-s)}\left[\left(\zeta_{i-p}\left(a_{i-p}-\dfrac{\zeta_{i-p}}{\xi_{i-p}}\right)-\xi_{i-p}b_{i-p}\right)\right.\right. \\
(\phi_{i-p}(s+t_n)-\phi_{i-p}(s)) \\
\left.\left.+\xi_{i-p}\displaystyle\sum_{j=1}^{p} c_{(i-p)j}(f_j(\phi_j(s+t_n))-f_j(\phi_j(s)))+\xi_{i-p}(v_{i-p}(s+t_n)-v_{i-p}(s))\right]d\tau ds\right| \\
+\left|\displaystyle\int_{c}^{t} e^{-\left(a_{i-p}-\frac{\zeta_{i-p}}{\xi_{i-p}}\right)(t-s)}\left[\left(\zeta_{i-p}\left(a_{i-p}-\dfrac{\zeta_{i-p}}{\xi_{i-p}}\right)-\xi_{i-p}b_{i-p}\right)\right.\right. \\
(\phi_{i-p}(s+t_n)-\phi_{i-p}(s)) \\
\left.\left.+\xi_{i-p}\displaystyle\sum_{j=1}^{p} c_{(i-p)j}(f_j(\phi_j(s+t_n))-f_j(\phi_j(s)))+\xi_{i-p}(v_{i-p}(s+t_n)-v_{i-p}(s))\right]ds\right| \\
\leqslant \dfrac{1}{a_{i-p}-\dfrac{\zeta_{i-p}}{\xi_{i-p}}}\left(2H\left|\zeta_{i-p}\left(a_{i-p}-\dfrac{\zeta_{i-p}}{\xi_{i-p}}\right)-\xi_{i-p}b_{i-p}\right|\right. \\
\left.+2LH\xi_{i-p}\displaystyle\sum_{j=1}^{p} c_{(i-p)j}+2H\xi_{i-p}\right)e^{-\left(a_{i-p}-\frac{\zeta_{i-p}}{\xi_{i-p}}\right)(a-c)} \\
+\dfrac{1}{a_{i-p}-\dfrac{\zeta_{i-p}}{\xi_{i-p}}}\left(\left|\zeta_{i-p}\left(a_{i-p}-\dfrac{\zeta_{i-p}}{\xi_{i-p}}\right)-\xi_{i-p}b_{i-p}\right|\xi+L\xi_{i-p}\displaystyle\sum_{j=1}^{p} c_{(i-p)j}\xi+\xi\xi_{i-p}\right) \\
<\epsilon, \\
i=p+1,\ldots,2p.
\end{cases} \tag{7.30}
$$

The inequalities (7.30) imply that $\|\Pi\phi(t+t_n)-\Pi\phi(t)\|<\epsilon$, for $t\in[a,b]$. Since ϵ is an arbitrarily small number, the condition (K3) is valid. The lemma is proved. □

Lemma 7.2.2. [16] *The operator* Π *is a contraction operator on* Σ.

Proof. Let us take two functions $u, v \in \Sigma$. For a fixed $t\in\mathbb{R}$ we obtain

$$|\Pi_i u_i(t) - \Pi_i v_i(t)| =$$

$$= \begin{cases} \left| \dfrac{1}{\xi_i} \displaystyle\int_{-\infty}^{t} e^{-\frac{\zeta_i}{\xi_i}(t-s)} (u_{i+p}(s) - v_{i+p}(s)) ds \right| \leqslant & i = 1, \ldots, p, \\ \dfrac{1}{\zeta_i} \|u(t) - v(t)\|_1, & \\ \left| \displaystyle\int_{-\infty}^{t} e^{-\left(a_{i-p} - \frac{\zeta_{i-p}}{\xi_{i-p}}\right)(t-s)} \left[\left(\zeta_{i-p} \left(a_{i-p} - \dfrac{\zeta_{i-p}}{\xi_{i-p}} \right) - \xi_{i-p} b_{i-p} \right) \right. \right. & \\ \qquad (u_{i-p}(s) - v_{i-p}(s)) & \\ \left. \left. + \, \xi_{i-p} \displaystyle\sum_{j=1}^{p} c_{(i-p)j} (f_{j-p}(u_{j-p}(s)) - f_{j-p}(v_{j-p}(s))) \right] ds \right| & \\ \leqslant \dfrac{1}{\left(a_{i-p} - \dfrac{\zeta_{i-p}}{\xi_{i-p}} \right)} \left[|\zeta_{i-p} \left(a_{i-p} - \dfrac{\zeta_{i-p}}{\xi_{i-p}} \right) - \xi_{i-p} b_{i-p}| \right. & \\ \left. + \, L \xi_{i-p} \displaystyle\sum_{j=1}^{p} c_{(i-p)j} \right] \|u(t) - v(t)\|_1, & i = p + 1, \ldots, 2p. \end{cases}$$

The conditions (C4) and (C7) imply that the operator Π is contractive in the set Σ.

Theorem 7.2.1. [16] *Assume that conditions (C1)–(C8) are fulfilled. Then the system* (7.18) *admits a unique asymptotically stable unpredictable solution.*

Proof. Let us show that the space Σ is complete. Consider a Cauchy sequence $\phi_k(t)$ in Σ, which converges to a limit function $\phi(t)$ on $\mathbb{R}$. It suffices to show that $\phi(t)$ satisfies condition (K3), since the other two conditions can easily be checked. Fix a closed and bounded interval $I \subset \mathbb{R}$. We have that

$$\begin{aligned} \|\phi(t + t_n) - \phi(t)\| &\leqslant \|\phi(t + t_n) - \phi_k(t + t_n)\| + \\ \|\phi_k(t + t_n) - \phi_k(t)\| &+ \|\phi_k(t) - \phi(t)\|. \end{aligned} \tag{7.31}$$

Now one can take sufficiently large n and k such that each term on the right-hand side of equation (7.31) is smaller than $\frac{\epsilon}{3}$ for an arbitrary positive ϵ and $t \in I$. The inequality implies that $\|\phi(t + t_n) - \phi(t)\| \leqslant \epsilon$ on I. That is, the sequence $\phi(t + t_n)$ uniformly converges to $\phi(t)$ on I. The completeness of space Σ are proved. By the contractive mapping theorem, due to lemmas 7.2.2 and 7.2.4, there exists a unique solution $\omega(t) \in \Sigma$ of equation (7.18).

Next we prove the unpredictability property. There exist a positive number κ and integers l and k such that the following inequalities are valid:

$$\kappa < \delta; \tag{7.32}$$

$$\xi_{i-p}\kappa \geqslant \left(\frac{1}{l} + \frac{2}{k}\right)\kappa\left[\left(a_{i-p} - \frac{\zeta_{i-p}}{\xi_{i-p}}\right) + |\zeta_{i-p}|\left(a_{i-p} - \frac{\zeta_{i-p}}{\beta}\right) - \beta b_{i-p}| + L\xi_{i-p}\sum_{j=1}^{p} c_{(i-p)j}\right] \tag{7.33}$$
$$+ \kappa\frac{\epsilon_0}{l} + \frac{1}{2l};$$

$$|\omega_{i-p}(t+s) - \omega_{i-p}(t)| < \epsilon_0 \min\left(\frac{1}{k}, \frac{1}{4l}\right), \quad t \in \mathbb{R}, |s| < \kappa, \tag{7.34}$$

for all $i = p + 1, \ldots, 2p$.

Let the numbers κ, l and k as well as numbers $n \in \mathbb{N}$, and $i = 1, \ldots, 2p$ be fixed. It is true that

$$\omega_i(t+t_n) - \omega_i(t) = \omega_i(s_n+t_n) - \omega_i(s_n) - \int_{s_n}^{t}\left(a_{i-p} - \frac{\zeta_{i-p}}{\xi_{i-p}}\right)(\omega_{i-p}(s+t_n) - \omega_{i-p}(s))ds$$
$$+ \int_{s_n}^{t}(F_i(\omega(s+t_n)) - F_i(\omega(s)))ds$$
$$= \omega_i(s_n+t_n) - \omega_i(s_n) - \int_{s_n}^{t}\left(a_{i-p} - \frac{\zeta_{i-p}}{\xi_{i-p}}\right)(\omega_{i-p}(s+t_n) - \omega_{i-p}(s))ds$$
$$- \int_{s_n}^{t}\left(\zeta_{i-p}\left(a_{i-p} - \frac{\zeta_{i-p}}{\xi_{i-p}}\right) - \xi_{i-p}b_{i-p}\right)(\omega_{i-p}(s+t_n) - \omega_{i-p}(s))ds$$
$$+ \int_{s_n}^{t}\xi_{i-p}\sum_{j=1}^{p} c_{(i-p)j}(f_j(\omega_j(s+t_n)) - f_j(\omega_j(s)))ds + \int_{s_n}^{t}\xi_{i-p}(v_{i-p}(s+t_n) - v_{i-p}(s))ds,$$

for $i = p + 1, \ldots, 2p$. Consider the following two alternatives: (i) $|\omega_i(t_n+s_n) - \omega_i(s_n)| < \epsilon_0/l$ and (ii) $|\omega_i(t_n+s_n) - \omega_i(s_n)| \geqslant \epsilon_0/l$ such that the remaining proof falls naturally into two parts.

(i) From equation (7.11) it follows that

$$\begin{aligned} |\omega_i(t+t_n) - \omega_i(t_n)| &\leqslant |\omega_i(t+t_n) - \omega_i(s_n+t_n)| + |\omega_i(s_n+t_n) - \omega_i(s_n)| \\ &+ |\omega_i(s_n) - \omega_i(t)| < \frac{\epsilon_0}{l} + \frac{\epsilon_0}{k} + \frac{\epsilon_0}{k} = \epsilon_0\left(\frac{1}{l} + \frac{2}{k}\right), \end{aligned} \tag{7.35}$$

if $t \in [s_n, s_n + \kappa]$.

Now, using relation (7.35), we have that

$$|\omega_i(t+t_n)-\omega_i(t)| \geqslant |\int_{s_n}^{t}\xi_{i-p}(v_{i-p}(s+t_n)-v_{i-p}(s))ds| - |\int_{s_n}^{t}\left(a_{i-p}-\frac{\zeta_{i-p}}{\xi_{i-p}}\right)(\omega_{i-p}(s+t_n)-$$
$$\omega_{i-p}(s))ds| - |\int_{s_n}^{t}\left(\zeta_{i-p}\left(a_{i-p}-\frac{\zeta_{i-p}}{\xi_{i-p}}\right)-\xi_{i-p}b_{i-p}\right)(\omega_{i-p}(s+t_n)-\omega_{i-p}(s))ds|$$
$$-|\int_{s_n}^{t}\xi_{i-p}\sum_{j=1}^{p}c_{(i-p)j}(f_j(\omega_j(s+t_n))-f_j(\omega_j(s)))ds| - |\int_{s_n}^{t}(\omega_{i-p}(s_n+t_n)-\omega_{i-p}(s_n))ds|$$
$$\geqslant \xi_{i-p}\epsilon_0\kappa - \epsilon_0\left(\frac{1}{l}+\frac{2}{k}\right)\kappa\left[\left(a_{i-p}-\frac{\zeta_{i-p}}{\xi_{i-p}}\right)+|\zeta_{i-p}\left(a_{i-p}-\frac{\zeta_{i-p}}{\xi_{i-p}}\right)-\xi_{i-p}b_{i-p}|+L\xi_{i-p}\sum_{j=1}^{p}c_{(i-p)j}\right]-$$
$$\kappa\frac{\epsilon_0}{l} \geqslant \frac{\epsilon_0}{2l},$$

for $t \in [s_n, s_n+\kappa]$ and $i = p+1, \dots, 2p$.

(ii) For the case $\Delta \geqslant \epsilon_0/l$, from equation (7.11) we obtain that

$$|\omega_i(t+t_n)-\omega_i(t)| \geqslant |\omega_i(t_n+s_n)-\omega_i(s_n)| - |\omega_i(s_n)-\omega_i(t)| - |\omega(t+t_n)-\omega(t_n+s_n)|$$
$$\geqslant \frac{\epsilon_0}{l}-\frac{\epsilon_0}{4l}-\frac{\epsilon_0}{4l}=\frac{\epsilon_0}{2l}, \tag{7.36}$$

if $t \in [s_n-\kappa, s_n+\kappa]$ and $n \in \mathbb{N}$. Thus we obtain that $\|\omega(t+t_n)-\omega(t)\| \geqslant \epsilon_0$ for $t \in [s_n, s_n+\kappa]$. That is, one can conclude that $\omega(t)$ is an unpredictable solution.

Now we will discuss the stability of the solution $\omega(t)$. It is true that

$$\omega(t) = e^{A(t-t_0)}\omega(t_0) + \int_{t_0}^{t} e^{A(t-s)}F(\omega(s))ds.$$

Let $\psi(t) = (\psi_1, \psi_2, \dots, \psi_{2p})$ be another solution of system (7.18). One can write

$$\psi(t) = e^{A(t-t_0)}\psi(t_0) + \int_{t_0}^{t} e^{A(t-s)}F(\psi(s))ds.$$

We denote $\lambda = \min\left(\frac{\zeta_i}{\xi_i}; a_i - \frac{\zeta_i}{\xi_i}\right)$ and $L_F = \max\left(\frac{1}{\xi_i}; |-\xi_i b_i + \zeta_i\left(a_i - \frac{\zeta_i}{\xi_i}\right)| + \xi_i L\sum_{j=1}^{p}c_{ij}\right)$.

Then we have that

$$\|\omega(t)-\psi(t)\| \leqslant e^{-\lambda(t-t_0)}\|\omega(t_0)-\psi(t_0)\| + \int_{t_0}^{t} e^{-\lambda(t-s)}L_F\|\omega(s))-\psi(s)\|ds, \ t \geqslant t_0.$$

Applying the Gronwall–Bellman lemma, one can obtain that

$$\|\omega(t)-\psi(t)| \leqslant \|\omega(t_0)-\psi(t_0)\|e^{(L_F-\lambda)(t-t_0)}, \ t \geqslant t_0,$$

and condition (C8) implies that $\omega(t)$ is a uniformly exponentially stable solution of the system (7.18). The theorem is proved. □

We shall need the following lemmas for the example.

Lemma 7.2.3. [12] *If the function $\phi(t)$: $\mathbb{R} \to \mathbb{R}$ is unpredictable then the function $\phi(t) + C$, where C is a constant, is also unpredictable.*

Lemma 7.2.4. [12] *Suppose that $\phi(t)$: $\mathbb{R} \to \mathbb{R}$ is an unpredictable function. Then the function $\phi^3(t)$ is unpredictable.*

Example 7.2.1. [16] *According to example* 2.2.1, *the logistic map*

$$\lambda_{i+1} = F_\mu(\lambda_i), \tag{7.37}$$

where $i \in \mathbb{Z}$ and $F_\mu(s) = \mu s(1 - s)$, has an unpredictable solution ϕ_i. Let $\Omega(t)$ be a piecewise constant function which is defined by $\Omega(t) = \psi_i$ for $t \in [i, i + 1), i \in \mathbb{Z}$. As a function of external inputs, we will use an unpredictable function of example 7.1.10,

$$\Theta(t) = \int_{-\infty}^{t} e^{-3(t-s)}\Omega(s)ds. \tag{7.38}$$

Let us take into account the system

$$\frac{d^2x_i(t)}{dt^2} = -a_i\frac{x_i(t)}{dt} - b_ix_i(t) + \sum_{j=1}^{3} c_{ij}f_j(x_j(t)) + v_i(t), \tag{7.39}$$

where $i = 1, 2, 3$ $a_1 = 6,$ $a_2 = 5,$ $a_3 = 7,$ $b_1 = 8,$ $b_2 = 6,$ $b_3 = 8,$ $f(x) = 0.35\arctan(x)$,

$$\begin{pmatrix} c_{11} & c_{12} & c_{13} \\ c_{21} & c_{22} & c_{23} \\ c_{31} & c_{32} & c_{33} \end{pmatrix} = \begin{pmatrix} 0.02 & 0.03 & 0.02 \\ 0.04 & 0.05 & 0.01 \\ 0.03 & 0.06 & 0.02 \end{pmatrix},$$

and $v_1(t) = -58\Theta^3(t) + 5$, $v_2(t) = 76\Theta^3 + 4$ and $v_3(t) = 42\Theta(t) - 3$. *According to lemmas* 7.2.8 *and* 7.2.9 *the function $v(t) = (v_1(t), v_2(t), v_3(t))$ is strongly unpredictable. The conditions* (C1) – (C8) *hold for the network* (7.39) *with* $\xi_1 = \xi_2 = 2$, $\xi_3 = 3$, $\zeta_1 = \zeta_2 = 4$, $\zeta_3 = 7$, $L = 0.35$, $M_f = 0.56$ and $H = 17.1$. *In figures* 7.4 *and* 7.5 *we depict the coordinates and trajectory of the solution $\omega(t)$ with initial values* $\omega_1(0) = 1.023$, $\omega_2(0) = 1.516$ and $\omega_3(0) = 0.275$. *The solution $\omega(t)$ approximates the asymptotically stable unpredictable solution of the equation* (7.39) *as time increases.*

7.3 Notes

Recent developments in the field of neural networks have led to increased interest in the complexity of the dynamics. Oscillations and chaos in neural networks are real and have stimulated the interest of many scientists [23, 28, 43, 56, 66]. They occur in a neural network system due to properties of single neurons [23, 52, 56] and synaptic connections among neurons [20, 53]. The neural networks of the current research display unpredictable oscillations and chaos. The unpredictable function is introduced in [4] and is based on the dynamics of unpredictable points and Poincaré chaos [5].

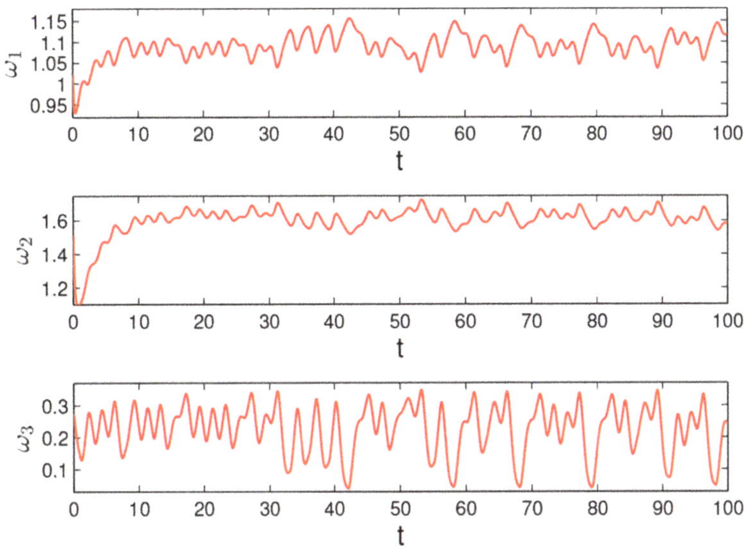

Figure 7.4. The coordinates of the function $\omega(t)$, which exponentially converges to the coordinates of the strongly unpredictable solution of equation (7.39).

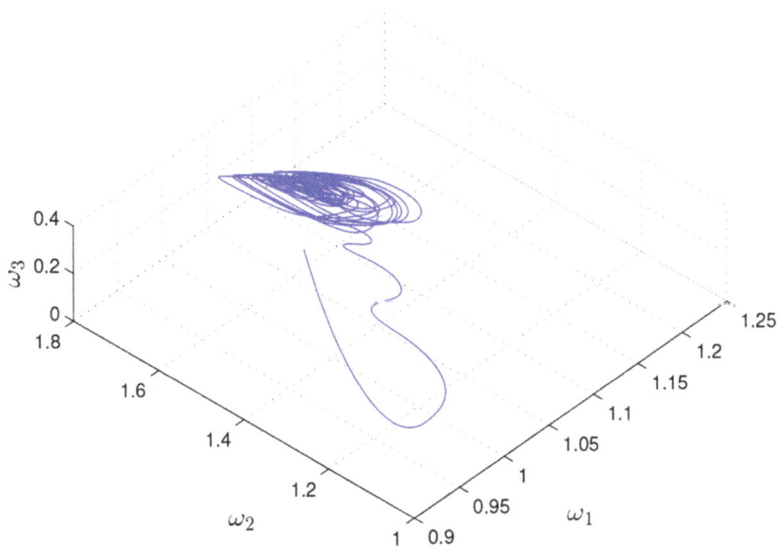

Figure 7.5. The trajectory of the function $\omega(t)$.

More precisely, the function is an unpredictable point of the Bebutov dynamics and consequently it is a member of the chaotic set [6]. The notion of the unpredictable point extends the frontiers of the classical theory of dynamical systems and the unpredictable function provides new problems of the existence of unpredictable oscillations for the theory of differential equations. These studies have been identified as major contributing factors for the emergence of new types of sophisticated motion. Significant results have been obtained for unpredictable oscillations of Hopfield type neural networks, shunting inhibitory cellular neural networks and inertial neural networks [3, 13–16].

References

[1] Aihara K, Takabe T and Toyoda M 1990 Chaotic neural networks *Phys. Lett.* A **144** 333–40

[2] Akhmet M, Aruğaslan D and Yilmaz E 2010 Stability analysis of recurrent neural networks with piecewise constant argument of generalized type *Neural Netw.* **23** 805–11

[3] Akhmet M, Aruğaslan Çinçin D, Tleubergenova M, Nugayeva Z, Tleubergenova M and Nugayeva Z 2021 Unpredictable oscillations for Hopfield-type neural networks with delayed and advanced arguments *Mathematics* **9** 1–19

[4] Akhmet M and Fen M O 2016 Poincaré chaos and unpredictable functions *Commun. Nonlinear Sci. Numer. Simul.* **48** 85–94

[5] Akhmet M and Fen M O 2016 Unpredictable points and chaos *Commun. Nonlinear Sci. Numer. Simul.* **40** 1–5

[6] Akhmet M and Fen M O 2017 Existence of unpredictable solutions and chaos *Turk. J. Math.* **41** 254–66

[7] Akhmet M and Fen M O 2018 Non-autonomous equations with unpredictable solutions *Commun. Nonlinear Sci. Numer. Simul.* **59** 657–70

[8] Akhmet M, Fen M O and Alejaily E M 2018 Dynamics motivated by Sierpinski fractals arXiv: 1811.07122

[9] Akhmet M and Fen M O 2015 Attraction of Li–Yorke chaos by retarded SICNNs *Neurocomputing* **147** 330–42

[10] Akhmet M, Fen M O and Kirane M 2015 Almost periodic solutions of retarded SICNNs with functional response on piecewise constant argument *Neural Comput. Appl.* **13** 1–10

[11] Akhmet M, Fen M O, Tleubergenova M and Zhamanshin A 2019 Poincaré chaos for a hyperbolic quasilinear system *Miskolc Math. Notes* **20** 33–44

[12] Akhmet M, Fen M O, Tleubergenova M and Zhamanshin A 2019 Unpredictable solutions of linear differential and discrete equations *Turk. J. Math.* **43** 2377–89

[13] Akhmet M, Seilova R, Tleubergenova M and Zhamanshin A 2020 Shunting inhibitory cellular neural networks with strongly unpredictable oscillations *Commun. Nonlinear Sci. Numer. Simul.* **89** 105287

[14] Akhmet M, Tleubergenova M and Nugayeva Z 2020 Strongly unpredictable oscillations of Hopfield-type neural networks *Mathematics* **8** 1791

[15] Akhmet M, Tleubergenova M and Zhamanshin A 2019 Neural networks with Poincaré chaos *ACM Int. Conf. Proc. Series* pp 34–7

[16] Akhmet M, Tleubergenova M and Zhamanshin A 2020 Inertial neural networks with unpredictable oscillations *Mathematics* **8** 1797

[17] Akhmet M and Yilmaz E 2013 *Neural Networks with Discontinuous/Impact Activations* (New York: Springer)

[18] Arbib M A 1988 *Brain Theory and Neural Networks* (Cambridge, MA: MIT Press)

[19] Babcock K L and Westervelt R M 1986 Stability and dynamics of simple electronic neural networks with added inertia *Physica* D **23** 464–9

[20] Bel A and Rotstein H G 2019 Membrane potential resonance in non-oscillatory neurons interacts with synaptic connectivity to produce network oscillations *J. Comput. Neurosci.* **46** 169–95

[21] Bouzerdoum A and Pinter R B 1993 Shunting inhibitory cellular neural networks: derivation and stability analysis *IEEE Trans. Circ. Syst. I Fund. Theory Appl.* **40** 215–21
[22] Chen L and Zhao H 2008 Global stability of almost periodic solution of shunting inhibitory cellular neural networks with variable coefficients *Chaos Soliton. Fract.* **35** 351–7
[23] Derrida B and Meir R 1988 Chaotic behavior of a layered neural network *Phys. Rev.* A **38** 3116–9
[24] Ding H S, Liang J and Xiao T J 2008 Existence of almost periodic solutions for SICNNs with time-varying delays *Phys. Lett.* A **372** 5411–6
[25] Dong Q, Matsui K and Huang X 2002 Existence and stability of periodic solutions for Hopfield neural network equations with periodic input *Nonlinear Anal.* **49** 471–9
[26] Driver R D 1977 *Ordinary and Delay Differential Equations* vol 20 (New York: Springer)
[27] Gray C M, König P, Engel A K and Singer W 1989 Oscillatory responses in cat visual cortex exhibit inter-columnar synchronization which reflects global stimulus properties *Nature* **338** 334–7
[28] Guevara M R, Glass L, Mackey M C and Shrier A 1983 Chaos in neurobiology *IEEE Trans. Syst. Man Cybern.* **13** 790–8
[29] Gui Z 2006 Periodic solution and chaotic strange attractor for shunting inhibitory cellular neural networks with impulses *Chaos* **16** 033116
[30] Hartman P 2002 *Ordinary Differential Equations* (New York: SIAM)
[31] Hayashi Y 1994 Oscillatory neural network and learning of continuously transformed patterns *Neural Netw.* **7** 219–31
[32] He G, Shrimali M D and Aihara K 2008 Threshold control of chaotic neural network *Neural Netw.* **21** 114–21
[33] He Y, Wu M and She J-H 2006 Delay-dependent exponential stability of delayed neural networks with time-varying delay *EEE Trans. Circuits Syst. II Exp. Briefs* **53** 553–7
[34] Hoppensteadt F C and Izhekevich E M 1999 Oscillatory neurocomputers with dynamic connectivity *Phys. Rev. Lett.* **82** 2983–6
[35] Huang X and Cao J 2003 Almost periodic solution of shunting inhibitory cellular neural networks with time-varying delay *Phys. Lett.* A **314** 222–31
[36] Hutcheon B and Yarom Y 2000 Resonance, oscillation and the intrinsic frequency preferences of neurons *Trends Neurosci.* **23** 216–22
[37] Jiang H, Zhang L and Teng Z 2005 Existence and global exponential stability of almost periodic solution for cellular neural networks with variable coefficients and time-varying delays *IEEE Trans. Neural Netw.* **16** 1340–51
[38] Ge J and Hu J 2012 Weak resonant double Hopf bifurcations in an inertial four-neuron model with time delay *Int. J. Neural Syst.* **22** 63–75
[39] Ke Y Q and Miao C F 2011 Stability analysis of BAM neural networks with inertial term and time delay *WSEAS Trans. Syst.* **10** 425–38
[40] Ke Y Q and Miao C F 2013 Stability analysis of inertial Cohen–Grossberg-type neural networks with time delays *Neurocomputing* **117** 196–205
[41] Ke Y Q and Miao C F 2013 Stability and existence of periodic solutions in inertial BAM neural networks with time delay *Neural Comput. Appl.* **23** 1089–99
[42] Korn H and Faure P 2003 Is there chaos in the brain? II. Experimental evidence and related models *Neurosciences, C. R. Biol.* **326** 787–840
[43] Landau I D and Sompolinsky H 2018 Coherent chaos in a recurrent neural network with structured connectivity *PLoS Comput. Biol.* **14** e1006309

[44] Li R and Zhou X 2013 A note on chaos in product maps *Turk. J. Math.* **37** 665–75
[45] Li X and Song S 2013 Impulsive control for existence, uniqueness, and global stability of periodic solutions of recurrent neural networks with discrete and continuously distributed delays *IEEE Trans. Neural. Netw. Learn. Syst.* **24** 868–77
[46] Li Y and Meng X 2018 Almost periodic solutions for quaternion-valued shunting inhibitory cellular neural networks of neutral type with time delays in the leakage term *Int. J. Syst. Sci.* **49** 2490–505
[47] Liao G, Chen X and Sanchez E N 2002 Delay-dependent exponential stability analysis of delayed neural networks: an LMI approach *Neural Netw.* **15** 855–66
[48] Liu B and Huang L 2008 Positive almost periodic solutions for recurrent neural networks *Nonlinear Anal.: Real World Appl.* **9** 830–41
[49] Liu Z and Liao L 2004 Existence and global exponential stability of periodic solution of cellular neural networks with time-varying delays *J. Math. Anal. Appl.* **290** 247–62
[50] Llinas R 1988 The intrinsic electrophysiological properties of mammalian neurons: insights into central nervous system function *Science* **242** 1654–64
[51] Miller V H and Jansen B H 2008 Oscillatory neural network for pattern recognition: trajectory based classification and supervised learning *Biol. Cybern.* **99** 459–71
[52] Muscinelli S P, Gerstner W and Schwalger T 2019 How single neuron properties shape chaotic dynamics and signal transmission in random neural networks *PLoS Comput. Biol.* **15** e1007122
[53] Penn Y, Segal M and Moses E 2016 Network synchronization in hippocampal neurons *PNAS* **113** 3341–6
[54] Qi J, Li C and Huang T 2015 Stability of inertial BAM neural network with time-varying delay via impulsive control *Neurocomputing* **161** 162–7
[55] Qin S, Gu L and Pan X 2020 Exponential stability of periodic solution for a memristor-based inertial neural network with time delays *Neural Comput. Appl.* **32** 3265–81
[56] Qu J, Wang R, Yan C and Du Y 2014 Oscillations and synchrony in a cortical neural network *Cogn. Neurodyn.* **8** 157–66
[57] Raiko T and Valpola H 2011 Oscillatory neural network for image segmentation with biased competition for attention *Adv. Exp. Med. Biol.* **718** 175–85
[58] Rakkiyappan R, Gayathri D, Velmurugan G and Cao J 2019 Exponential synchronization of inertial memristor-based neural networks with time delay using average impulsive interval approach *Neural Process. Lett.* **50** 2053–71
[59] Rakkiyappan R, Premalatha S, Chandrasekar A and Cao J 2016 Stability and synchronization analysis of inertial memristive neural networks with time delays *Cogn. Neurodyn.* **10** 437–51
[60] Schmidt H, Avitabile D, Montbrió E and Roxin A 2018 Network mechanisms underlying the role of oscillations in cognitive tasks *PLoS Comput. Biol.* **14** e1006430
[61] Selverston A I and Moulins M 1985 Oscillatory neural networks *Annu. Rev. Physiol.* **47** 29–48
[62] Skarda C A and Freeman W J 1987 How brains make chaos in order to make sense of the world? *Behav. Brain Sci.* **10** 161–95
[63] Thakur R and Das R 2020 Strongly Ruelle–Takens, strongly Auslander–Yorke and Poincaré chaos on semiflows *Commun. Nonlinear Sci. Numer. Simul.* **81** 105018
[64] Wanatabe M, Aihara K and Kondo S 1997 Self-organization dynamics in chaotic neural networks *Control Chaos Model* **8** 320–33

[65] Wang L 2010 Existence and global attractivity of almost periodic solutions for delayed high-ordered neural networks *Neurocomputing* **73** 802–8
[66] Wang L, Pichler E E and Ross J 1990 Oscillations and chaos in neural networks: an exactly solvable model *Proc. Natl Acad. Sci.* 87 9467–71
[67] Wen S, Huang T, Zeng Z, Chen Y and Li P 2015 Circuit design and exponential stabilization of memristive neural networks *Neural Netw.* **63** 48–56
[68] Winfree A 1980 *The Geometry of Biological Time* (New York: Springer)
[69] Xia Y, Cao J and Huang Z 2007 Existence and exponential stability of almost periodic solution for shunting inhibitory cellular neural networks with impulses *Chaos Soliton. Fract.* **34** 1599–607
[70] Yi Z, Heng P A and Fu A W C 1999 Estimate of exponential convergence rate and exponential stability for neural networks *IEEE Trans. Neural Netw.* **10** 1487–93
[71] Yu B and Liu W 2004 Fractal analysis of permeabilities for porous media *AIChE J.* **50** 46–57
[72] Zhang W, Huang T, Li C and Yang J 2018 Robust stability of inertial BAM neural networks with time delays and uncertainties via impulsive effect *Neural Process. Lett.* **48** 245–56
[73] Zhang Z and Quan Z 2015 Global exponential stability via inequality technique for inertial BAM neural networks with time delays *Neurocomputing* **151** 1316–26
[74] Zhao Y, Cai Y and Fan G 2016 Dynamical behavior for fractional-order shunting inhibitory cellular neural networks *J. Nonlinear Sci. Appl.* **9** 4589–99
[75] Zhou Q and Zhao C 2010 Exponential convergence of shunting inhibitory cellular neural networks with continuously distributed delays *Ann. Pol. Math.* **98** 241–50

Lightning Source UK Ltd.
Milton Keynes UK
UKHW051615250921
391114UK00003B/76

9 780750 335058